Molecular Sieves

Science and Technology

Editors: H. G. Karge · J. Weitkamp

Volume 2

Springer-Verlag
Berlin Heidelberg GmbH

Structures and Structure Determination

With contributions by
C. Baerlocher, J. M. Bennett, W. Depmeier,
A. N. Fitch, H. Jobic, H. van Koningsveld,
W. M. Meier, A. Pfenninger, O. Terasaki

Springer

Editors:

Dr. Hellmut G. Karge
Fritz Haber Institute of the Max Planck Society
Faradayweg 4–6
D-14195 Berlin
Germany

Professor Dr.-Ing. Jens Weitkamp
Institute of Chemical Technology I
University of Stuttgart
D-70550 Stuttgart
Germany

ISBN 978-3-662-30834-9

Library of Congress Cataloging-in-Publication Data

Molecular sieves : science and technology / editors. H. G. Karge, J. Weitkamp.
p. cm. Includes bibliographical references and index. Contents: v. 1. Synthesis
ISBN 978-3-662-30834-9 ISBN 978-3-540-69749-7 (eBook)
DOI 10.1007/978-3-540-69749-7

1. Molecular sieves. I. Karge, H. G. (Hellmut G.) II. Weitkamp, J. (Jens)
TP159.M6M63 1998 660′.2842–dc21 98-15578

Originally published by Springer-Verlag Berlin Heidelberg New York in 1999
Softcover reprint of the hardcover 1st edition 1999

Typesetting: Dataconversion by Fotosatz-Service Köhler GmbH, Würzburg
Coverdesign: Friedhelm Steinen-Broo, Estudio Calamar, Pau/Spain;
design & production, Heidelberg

SPIN: 10502020 2/3020 – 5 4 3 2 1 0 – Printed on acid-free paper

Preface to the Series

Following Springer's successful series *Catalysis – Science and Technology*, this series of monographs has been entitled *Molecular Sieves – Science and Technology*. It will cover, in a comprehensive manner, all aspects of the science and application of zeolites and related microporous and mesoporous materials.

After about 50 years of prosperous research, molecular sieves have gained a firm and important position in modern materials science, and we are witnessing an ever increasing number of industrial applications. In addition to the more traditional and still prevailing applications of zeolites as water softeners in laundry detergents, as adsorbents for drying, purification and separation purposes, and as catalysts in the petroleum refining, petrochemical and chemical industries, novel uses of molecular sieves are being sought in numerous laboratories.

By the beginning of 1999, the Structure Commission of the International Zeolite Association had approved approximately 120 different zeolite structures which, altogether, cover the span of pore diameters from about 0.3 nm to 2 nm. The dimensions of virtually all molecules (except macromolecules) chemists are concerned with fall into this same range. It is this coincidence of molecular dimensions and pore widths which makes zeolites so unique in adsorption and catalysis and enables molecular sieving and shape-selective catalysis. Bearing in mind that each zeolite structure can be modified by a plethora of post-synthesis techniques, an almost infinite variety of molecular sieve materials are nowadays at the researcher's and engineer's disposal. In many instances this will allow the properties of a zeolite to be tailored to a desired application. Likewise, remarkable progress has been made in the characterization of molecular sieve materials by spectroscopic and other physico-chemical techniques, and this is particularly true for structure determination. During the last decade, we have seen impressive progress in the application of quantum mechanical ab initio and other theoretical methods to zeolite science. The results enable us to obtain a deeper understanding of physical and chemical properties of zeolites and may render possible reliable predictions of their behavior. All in all, the science and application of zeolites is a flourishing and exciting field of interdisciplinary research which has reached a high level of sophistication and a certain degree of maturity.

The editors believe that, at the turn of the century, the time has come to collect and present the huge knowledge on zeolite molecular sieves. *Molecular Sieves – Science and Technology* is meant as a handbook of zeolites, and the term "zeolites" is to be understood in the broadest sense of the word. While, throughout the handbook, some emphasis will be placed on the more traditional alumosilicate zeolites with eight-, ten- and twelve-membered ring pore openings, materials with other chemical compositions and narrower and larger pores (such as sodalite, clathrasils, $AlPO_4$–8, VPI-5 or cloverite) will be covered as well. Also included are microporous forms of silica (e.g., silicalite-1 or -2), alumophosphates, gallophosphates, silicoalumophosphates and titaniumsilicalites etc. Finally, zeolite-like amorphous mesoporous materials with ordered pore systems, especially those belonging to the M41S series, will be covered. Among other topics related to the science and application of molecular sieves, the book series will put emphasis on such important items as: the preparation of zeolites by hydrothermal synthesis; zeolite structures and methods for structure determination; post-synthesis modification by, e.g., ion exchange, dealumination or chemical vapor deposition; the characterization by all kinds of physico-chemical and chemical techniques; the acidic and basic properties of molecular sieves; their hydrophilic or hydrophobic surface properties; theory and modelling; sorption and diffusion in microporous and mesoporous materials; host/guest interactions; zeolites as detergent builders; separation and purification processes using molecular sieve adsorbents; zeolites as catalysts in petroleum refining, in petrochemical processes and in the manufacture of organic chemicals; zeolites in environmental protection; novel applications of molecular sieve materials.

The handbook will appear over several years with a total of ten to fifteen volumes. Each volume of the series will be devoted to a specific sub-field of the fundamentals or application of molecular sieve materials and contain five to ten articles authored by renowned experts upon invitation by the editors. These articles are meant to present the state of the art from a scientific and, where applicable, from an industrial point of view, to discuss critical pivotal issues and to outline future directions of research and development in this sub-field. To this end, the series is intended as an up-to-date highly sophisticated collection of information for those who have already been dealing with zeolites in industry or at academic institutions. Moreover, by emphasizing the description and critical assessment of experimental techniques which have been used in molecular sieve science, the series is also meant as a guide for newcomers, enabling them to collect reliable and relevant experimental data.

The editors would like to take this opportunity to express their sincere gratitude to the authors who spent much time and great effort on their chapters. It is our hope that *Molecular Sieves – Science and Technology* turns out to be both a valuable handbook the advanced researcher will regularly consult and a useful guide for newcomers to the fascinating world of microporous and mesoporous materials.

Hellmut G. Karge
Jens Weitkamp

Preface to Volume 2

Once a new natural zeolite is found or a new molecular sieve synthezised, via one or the other of the methods described in Volume 1 for example, the researchers face the task of confirming that a novel structure has come into their hands. However, beyond this basic problem, questions soon arise concerning rather detailed and subtle structural features.

The classical method of determining crystal structures is X-ray diffraction. Thus, in Chapter 1 of the present volume, *H. van Koningsveld* and *M. Bennett* provide the reader with information about the enormous progress which has been made in X-ray structure analysis of zeolites. To a large extent, this is due to outstanding developments in both experimental techniques and methods of data evaluation, such as the application of synchrotron radiation and Rietveld analysis. New methods now enable crystallographers to study very small single crystals or crystallite powders. This is extremely important with respect to most of the synthetic micro- and mesoporous materials since the size of primary particles is usually in the μm range. The authors stress that, in the context of reliable structure analysis, the determination of the unit cell and space group is of paramount importance. Modern tools now allow researchers to study subtle effects on zeolite structures such as those caused by framework distortions, dealumination, isomorphous substitution or cation and sorbate location.

The study of structures containing light atoms is the particular domain of neutron scattering, even though this is not its only advantage. The authors of Chapter 2, *A. N. Fitch* and *H. Jobic* demonstrate the way in which neutron scattering is able to complement structure analysis by X-ray diffraction. In particular, neutron scattering techniques reveal their strong potential in probing details of structural arrangements involving hydrogen-containing species (such as water and hydroxyl groups) as well as determining hydrogen bonds, cation positions, and the location of adsorbed molecules. Frequently these techniques are successfully used for further refinement of X-ray diffraction data.

Chapter 3, written by *O. Terasaki*, is devoted to the use of the various kinds of electron microscopy in the investigation of zeolites and related porous solids. The author's contribution focuses on the potential of electron microscopy in studying crystallite morphologies as well as features of the fine structure, e.g., bulk and surface defects; details of the crystal surface (edges and kinks), and, as such, related to crystal growth; and modification of frameworks. Moreover, the valuable assistance of electron microscopy in solving new structures is illustrated by a number of examples.

Chapter 4 is contributed by *W. Depmeier*, and it concerns particular phenomena of the structures of zeolites and related solids which are attracting more and more interest. Such phenomena are, inter alia, phase transitions as well as mechanisms of reduction in symmetry and volume as a consequence of tilting, distortion of the whole framework or framework units, modulations of the framework, and partial amorphization. These are demonstrated by a variety of instructive examples, and their importance is pointed out in view of, for example, catalytic, shape selective and separation properties of zeolite materials.

General problems of zeolite structures are dealt with in Chapter 5 which is jointly authored by *W.M. Meier* and *C. Baerlocher*. It includes basic aspects of zeolite crystallography such as topology, configuration, and conformation of framework structures. Similarly, the idea of distinguishing zeolites on the basis of framework densities is presented. The attempts at classification of zeolite structure types are critically discussed. The authors then describe the interesting concepts of structural characterization via loop configurations and coordination sequences and also reconsider the long-standing question of whether zeolite framework structures are predictable.

This volume concludes with Chapter 6, a review devoted to industrial synthesis. Contributed by *A. Pfenninger* and entitled "Manufacture and Use of Zeolites for Adsorption Processes", this chapter provides an extremely useful adjunct to Volume 1 of this series. Important aspects of industrial synthesis are described and, simultaneously, the characterization and use of zeolites for separation processes are discussed. In these respects, Chapter 6 is something of an introduction to matters which will be extensively dealt with in Volume 5 (Characterization II) and Volume 7 (Sorption and Diffusion) of this series.

The originally planned final chapter on the role played by solid state NMR spectroscopy in the elucidation of structural features of microporous and mesoporous materials was unfortunately not available at the time of going to press. However, given the importance of this topic, an appropriate treatment of this area is intended to appear in Volume 4 (Characterization I).

Thus, Volume 2 presents an extended overview over most of the relevant techniques currently employed for investigations into structural properties of micro- and mesoporous materials and offers in its last contribution a valuable addition to the topics treated in Volume 1. From this volume it becomes evident that the various techniques for structure determination are, to a large extent, complementary and that evaluation of the experimental data, on the other hand, is profiting much from recent developments in theory and modeling. It is the Editors' hope that Volume 2 of the series "Molecular Sieves – Science and Technology" will provide the researchers in the field of zeolites and related materials with the necessary awareness of the great potential in modern methods for structure analysis.

Hellmut G. Karge
Jens Weitkamp

Contents

H. van Koningsveld and J.M. Bennett:
Zeolite Structure Determination from X-Ray Diffraction 1

A.N. Fitch and H. Jobic:
Structural Information from Neutron Diffraction 31

O. Terasaki:
Electron Microscopy Studies in Molecular Sieve Science 71

W. Depmeier:
Structural Distortions and Modulations in Microporous Materials 113

W.M. Meier and C. Baerlocher:
Zeolite Type Frameworks: Connectivities, Configurations and Conformations 141

A. Pfenninger:
Manufacture and Use of Zeolites for Adsorption Processes 163

Subject Index 199

Author Index Vols. 1 and 2 215

Zeolite Structure Determination from X-Ray Diffraction

H. van Koningsveld[1] and J. M. Bennett[2]

[1] Laboratory of Organic Chemistry and Catalysis, Delft University of Technology, Julianalaan 136, 2628 BL Delft, The Netherlands; *e-mail:* havank@cad4sun.tn.tudelft.nl
[2] 661 Weadley Road, Radnor, PA 19087, USA; *e-mail:* JMBXrayse@aol.com

1 Introduction . . . 1

2 Severe Overlap of Reflections in Powder Data . . . 3

3 Incorrect Determination of the Space Group . . . 5

4 Effect of Framework Flexibility . . . 8

5 Disorder of Non-Framework Species . . . 16

6 Faulting within the Framework . . . 23

7 Isomorphous Replacement of Framework Atoms . . . 24

8 Crystal Size Limitations . . . 25

9 Conclusions . . . 25

References . . . 26

1 Introduction

Zeolites and related microporous materials are a class of materials with an ever widening range of compositions, structures and uses. Since the earliest days of zeolite science X-ray diffraction has been one of the basic and most useful tools for characterization.

Initially X-ray diffraction was used to answer simple questions such as: "have I made a new material?" or: "has the crystallization process gone to completion?" Now the questions encompass everything that a researcher might want to know about the structure of a material. Early attempts at determining crystal structures using X-ray diffraction were often unsuccessful because many of these early synthetic materials were available only as powder samples. Fortunately many of these first synthetic materials had natural counterparts with large single crystals, and data from these were used to determine the framework structures

Molecular Sieves, Vol. 2

of their synthetic counterparts. Today, the framework of a new material can be often determined from powder samples. In addition, single crystal techniques have improved considerably leading to increased accuracy in the bond angles and bond distances and to the ability to study crystals of much smaller size. It is now possible for a single crystal study to reveal details of the structure that show the interaction of a sorbed material with the framework or movement of cations within the framework and any ensuing distortions of the framework. Structural data from powder samples are beginning to reveal similar changes in the crystal structure with temperature, with sorbed materials and even under catalytic conditions. Even though the technique of X-ray powder diffraction has improved greatly since the early days of zeolite science, it is still more accurate to determine the crystal structure of a new material from single crystal data rather than from powder data.

Many of the advances in the structural information derived for zeolitic materials are a direct result of major improvements in powder and single crystal X-ray equipment available, in the development of new structure determination methods and in the use of new characterization tools including magic angle spinning NMR, neutron diffraction and electron microscopy, which are described in subsequent chapters. Two excellent review papers [1, 2] discuss the use of X-ray diffraction techniques to study zeolites and the problems encountered, and it is recommended that they be used in combination with this chapter.

The stages in determining the crystal structure of a material have been described as: (i) obtain a suitable sample, (ii) collect the data, (iii) determine a trial structure using *ab initio* methods, and (iv) refine the data.

However, with zeolites it is not as simple as the above infers since subtle changes in the zeolite framework can influence, to a greater or lesser extent, both the observed intensities and the symmetry. These subtle changes in the observed intensities and the symmetry can cause serious problems for crystallographers performing a zeolite structure analysis. The crystallographic problems include:

- Severe overlap of reflections in powder data leading to problems with the techniques used to decompose the peaks into individual reflections
- Incorrect determination of the space group especially when the true symmetry is masked by pseudo-symmetry
- The effect of framework flexibility on the structure analysis
- Disorder of the non-framework species and its effect on the structure solution
- Faulting within the framework
- Problems caused by isomorphous replacement of framework atoms
- The effects due to small crystal size and the limits on the crystal size that can be used

In order to help those in the zeolite community to better appreciate the beauty of an excellent crystallographic study while learning to evaluate the pitfalls that are present in an incorrect study, several structures, published in the last decade and that are examples of the problems listed above, will be reviewed.

2
Severe Overlap of Reflections in Powder Data

For a single crystal structure determination one crystal is chosen from the sample and it is assumed that the chosen crystal is both suitable for the study and typical of the bulk material. Often several crystals have to be evaluated before a "good" crystal for the study is found. In contrast, it is relatively easy to obtain a sample for a powder study and to use a synchrotron source to obtain the best data. Synchrotron X-ray data are high intensity and high resolution data and, as such, are far superior to in-house data. The improvements in the quality of the data obtained from the synchrotron have reduced the magnitude of the problems that plagued early attempts at structure determination. However, there is still only one dimensional intensity information in the powder pattern and it is not a trivial task to determine the correct three-dimensional unit cell dimensions especially if a few weak peaks from an unknown impurity phase are present.

A successful structure determination starts with a set of accurately determined peak positions. Unfortunately, this task is often left to the computer with disastrous results. With carefully deconvoluted data the currently used indexing programs [3, 4] often yield a number of equally probable answers. When combined with even partial unit cell information from electron diffraction, it is usually possible to reduce this number to one or two unit cell sets. If no other data are available then the wrong choice between two equally probable unit cells may prevent the structure from being accurately determined. Even when a unit cell is derived it may later prove to be "incorrect" (too highly symmetric) once the structure has been refined. Unfortunately, the only way to know that a chosen unit cell is correct is to solve the crystal structure.

Table 1 lists part of the data obtained from a new material. It was known from TEM/SEM studies that the synthesis product was impure and that the impurity was an offretite material based on observed d-spacings and a knowledge of the synthesis conditions. These offretite peaks were removed from the data before using the indexing programs. However, the best unit cell obtained did not index all the reflections suggesting that there might be three phases present in the sample which seemed unlikely. The final solution used several common reflections (such as that at $2\theta = 9.958°$) that came from both the offretite impurity and the new phase and indexed all 60 observed reflections, out to a d spacing of 3.04 Å. The only difference between the first and final unit cell solutions was the value for the c dimension. The number of un-indexed reflections now became zero (see Table 1). Thus it is very important to account for all observed peaks in a pattern even those assigned to other phases and to review even small differences between the observed and calculated 2θ values, in order to be sure that the calculated unit cell dimensions are reasonable.

In order to determine the crystal structure, the intensity of the exactly or partially overlapping reflections are usually separated by a number of simple techniques such as splitting them fifty-fifty. However, these structure determinations were often unsuccessful and more sophisticated methods were developed to partition the intensity of the overlapping reflections.

Table 1. A partial list of the observed and calculated 2θ values for a new phase with an offretite impurity[a]

2θ Values			Solution 1[b]	Final solution[c]
Observed	Calculated	Difference	h k l	h k l
2.402	2.396	0.007	1 0 0	1 0 0
4.142	4.150	0.008	1 1 0	1 1 0
4.784	4.792	0.009	2 0 0	2 0 0
6.606			1 0 0 (Off.)	1 0 0 (Off.)
6.338	6.341	0.003	2 1 0	2 1 0
7.189	7.191	0.002	3 0 0	3 0 0
8.302	8.305	0.003	2 2 0	2 2 0
9.595	9.593	-0.002	4 0 0	4 0 0
9.958	9.957	-0.001	0 0 1 (Off.)	0 0 1 (both)
10.244	10.243	-0.001	0 0 1	1 0 1
10.463	10.456	-0.007	3 2 0	3 2 0
10.793	10.791	-0.002	U	1 1 1
11.451			1 1 0 (Off.)	1 1 0 (Off.)
11.815	11.813	-0.002	U	2 1 1
12.002	11.999	-0.003	5 0 0	5 0 0
12.295	12.293	-0.002	U	3 0 1
12.478	12.471	-0.007	3 3 0	3 3 0
12.704	12.701	-0.003	4 2 0	4 2 0
12.980	12.980	0.000	U	2 2 1
13.369	13.367	-0.002	5 1 0	5 1 0
13.843	13.843	0.000	U	4 0 1

Off. indicates an offretite reflection and U an unindexed reflection
[a] Personal communication, Smith W, Bennett JM.
[b] Solution 1 had a = 36.147(3) and c = 7.329(1) Å.
[c] Final correct solution had a = 36.150(2) and c = 7.541(1) Å.

The use of Direct Methods in determining a weighting scheme for partitioning the intensities was developed by Jansen, Peschar and Schenk [5, 6]. The method was tested on a structure containing 22 atoms in the asymmetric unit cell; of the 527 observed reflections, 317 overlapped within half of the peak full width at half the peak maximum (FWHM) as determined in the fitting process [7]. Estermann et al. [8] described the structure determination of SAPO-40 (AFR)[1] using a different method for partitioning the intensities of the overlapping reflections. This Fast Iterative Patterson Squaring (FIPS) method indicated how to partition the intensity and only after this redistribution did an *ab initio* structure determination become possible.

Yet another method was applied to the structure determination of VPI-9 (VNI; [9]). This method uses a set of random starting phases for the intensities

[1] The arrangement of the tetrahedral atoms in most of the zeolite structures is indicated by a three letter code. This code is independent of the composition of the zeolite, the space group and symmetry. A full list of all currently assigned codes can be found in the 'Atlas of Zeolite Structure Types' by W.M. Meier, D.H. Olson and Ch. Baerlocher, Fourth Revised Edition, published on behalf of the Structure Commission of the International Zeolite Association by Elsevier, London, Boston, 1996.

obtained from the powder pattern and is then combined with a topological search routine in the Fourier recycling procedure. With this method both chemical and structural information are incorporated into the partitioning procedure used for the powder diffraction profile. With seven crystallographically unique tetrahedral sites, VPI-9 is the most complex framework arrangement currently solved from powder diffraction without manual intervention.

Since one-dimensional intensity data from powders is resolved into three-dimensional intensity data for single crystals, the problem with obtaining individual intensity data is not present with single crystal data. Therefore, the determination of the unit cell and symmetry is less difficult. Using the correct unit cell dimensions the intensities of all the single crystal reflections can be measured without serious overlap in most cases.

The lack of individually measured reflections with powder data also has a detrimental effect on the structure determination and refinement procedure. In powder diffraction the ratio between the number of observations and the number of parameters to be refined is very often less than or equal to one. However, with single crystal data this ratio usually ranges from three to ten. This over abundance of data allows an incomplete, or even partly wrong starting model to be used to yield a successful solution and final refinement of the structure. A recent example, illustrating the difference between powder and single crystal data, is the structure determination of $GaPO_4(OH)_{0.25}$ (-CLO; [10]). Even with high-resolution synchrotron powder data, 552 of the first 617 reflections have exact 2θ overlaps. This extreme example of the overlap of the individual intensity data could not be overcome until a large single crystal became available for conventional analysis. Then 2776 independent reflections were measured and the refinement converged smoothly.

3
Incorrect Determination of the Space Group

Space groups are determined from a list of hkl reflections that are not observed. This is very difficult with powder data because of the occurrence of overlapping reflections. Without a space group no crystal structure solution can be completed. However, in many cases it is not necessary to determine the space group that will result from a successful structure refinement. It is often only necessary to determine the starting space group that defines the maximum symmetry of the topology (maximum topological symmetry). For example, it is not necessary to differentiate between the tetrahedral aluminum and phosphorus atoms in a microporous aluminophosphate material in order to determine the correct framework topology. Fortunately, there have been found to be only a small number of maximum topology space groups that are applicable; some of them are *C2/m*, *Cmcm*, $I4_1/amd$ and $P6_3/mmc$. Since the choice of unit cell dimensions will affect the systematic absences and ultimately the space group, this knowledge of applicable space groups can be helpful when choosing between two different, but equally possible, unit cells. However, it must be remembered that the space group chosen must account for all of the low hkl systematic absences.

There are many different techniques used by crystallographers to arrive at the starting topology of a new material. All techniques, except model building, require that the space group be correctly determined. However, this very important step of determining the starting topology is often not adequately reported, possibly because it is the most time consuming step of a powder structure determination. It is possible to spend months to years determining the correct topology which, when determined, can lead to spending only days to weeks on the final refinement. The powder pattern of the proposed topology can be simulated after refinement of the interatomic distances using a Distance Least Squares (DLS) refinement [11] procedure and can then be compared to the experimental pattern of the material. Even when there is a passable match between the observed and simulated powder patterns it does not mean that the proposed framework arrangement is correct. Probably, any partially incorrect topology can be refined with the Rietveld technique [12] to yield an apparently acceptable solution.

ZSM-18 (MEI; [13]) is the only aluminosilicate zeolite that has been reported to contain a three tetrahedral atom ring (a T_3-ring)[2]. However, similar framework structures, such as MAPSO-46 (AFS; [14]), CoAPO-50 (AFY; [14]) and beryllophosphate-H (BPH; [15]), do not support this novel arrangement. An examination of the reported framework topology shows that the three ring arrangements can be replaced by a vertical SiOSi unit with practically no change in the positions of the remainder of the framework atoms. Lowering the symmetry by removing the six-fold axes and changing to orthorhombic symmetry allows the framework to rotate off the original six-fold axis thereby reducing the vertical SiOSi bond angles of 180°, which are undesirable but observed in the proposed structure. Unfortunately, any DLS refinement of an orthorhombic arrangement always refines back to a pseudo six-fold axis. The final answer to the question of whether ZSM-18 contains three rings will require a complete structure determination using powder data and consideration of the possibility that the original space group used to determine the structure was incorrect.

A postulated framework arrangement based on a DLS refinement should always be treated with suspicion because very few DLS refinements use the full symmetry of the chosen space group since the only symmetry operations needed are those that generate bonds that lie across the asymmetric unit cell boundaries. In addition, there is always the possibility that the space group chosen is incorrect and that therefore the final structure is incorrect as well. Several correct structures have been refined in two or more space groups and illustrate that there are subtle changes in the framework topology depending on the choice of space group [16].

An example showing that the observed distortions of the framework are dependent on the choice of the space group is given by the refinement of SAPO-40 (AFR; [17, 18]). The ordering of aluminum and phosphorus in the structure required that the c-axis be doubled and the space group be changed from orthorhombic *Pmmm* to monoclinic *P112/n*. Subsequently, it was realized that

[2] The standard method used to describe the number of atoms in a ring of a zeolite structure is to only count the tetrahedral (T) atoms. Thus a three ring opening would have three silicon atoms and the interconnecting three oxygen atoms for a total of six atoms.

this doubling generated c-glide planes and that the correct space group was actually orthorhombic *Pccn*. This change reduced the number of variables from 186 (for *P112/n*) to 95 (for *Pccn*) without affecting the quality of the profile fit. In addition many of the distances and angles, which were different in the *P112/n* refinement, become equivalent in *Pccn*. From a practical point of view it is very difficult to say which refinement yields a truer picture of the material and what effect the framework distortions will have on the material properties.

Sometimes the question of how material properties are affected by changes in the framework can be answered. In the case of VPI-5 (VFI; [19]) the recognition that octahedral aluminum is present in the structure of VPI-5 required that the symmetry be lowered from $P6_3cm$ to $P6_3$. Only after this symmetry change did the refinement of the structure proceed smoothly. The presence of a triple helix of occluded water molecules became evident because these water molecules were required to complete the octahedral coordination of half of the aluminum atoms in the fused 4-rings. The same octahedral coordination of aluminum was postulated for $AlPO_4H2$ (AHT; [20]), since both structures contain a triple crankshaft chain with fused 4-rings. Once the similar octahedral configuration was shown to be present, it was suggested that these octahedral distortions on the aluminum sites promote the reconstructive phase transition of VPI-5 to $AlPO_4$-8 (AET) above room temperature and of $AlPO_4H2$ to tridymite (Fig. 1) at higher temperatures. The phase transition of $AlPO_4H2$ to tridymite is irrever-

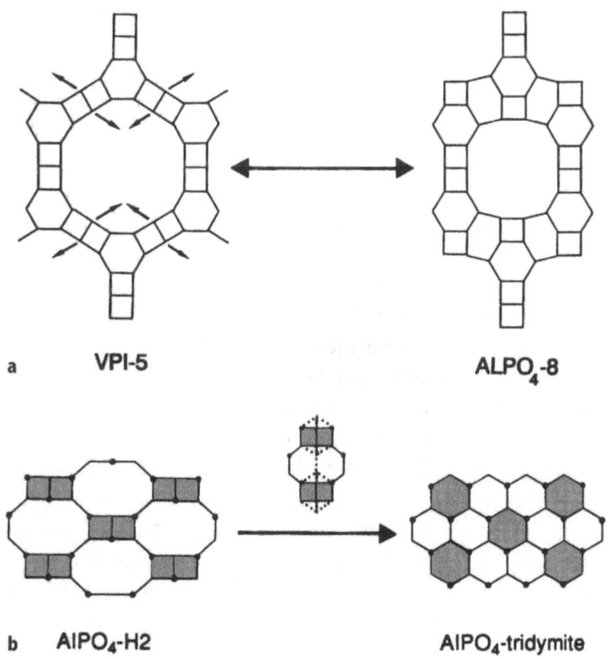

Fig. 1 a, b. Schematic illustration of the framework transformation of **a** VPI-5 to $AlPO_4$-8 and **b** $AlPO_4$-H2 to $AlPO_4$-tridymite. Large dots indicate Al positions. Reproduced by permission of the Royal Society of Chemistry from [20]

sible; conflicting reports exist as to whether the transition of VPI-5 to $AlPO_4$-8 is reversible or not [21–24].

4 Effect of Framework Flexibility

The difficulty in determining zeolite structures from diffraction data is increased when there are changes in cell dimensions and/or symmetry caused by the framework flexion in response to having different cations or other non-framework species present. The TO_4 tetrahedra are rigid but interconnected through oxygen atoms which act as flexible hinges [25]. In collapsible frameworks, such as ABW, GIS, NAT, RHO and SOD, all angles around the TO_4 tetrahedra co-rotate in the same sense when cell dimensions and volumes change. The frameworks wrap themselves around occluded non-framework species or collapse until the smallest angle of the TOT hinges (~126°) is reached [26]. MFI and MEL are also collapsible frameworks but do it through a shearing of the pentasil layers parallel to the crystallographic c axis [27]. In non-collapsible frameworks, such as LTA, FAU and KFI, the TOT hinges rotate in opposite directions when the cell dimensions and volume change. The frameworks are very flexible at intermediate values of the cell dimensions [28].

The distortions observed in the collapsible framework structures $MAlSiO_4$ (ABW) depend on the exchangeable cation M (where M is Li, Na, K, Rb, Cs, Tl or Ag) [29]. Even though it is theoretically possible to double the unit cell volume (Fig. 2), the space group usually remains $Pna2_1$ even with structures such as $LiGaAlSiO_4 \cdot H_2O$ [31], $LiBePO_4 \cdot H_2O$, $LiZnPO_4 \cdot H_2O$, $LiZnAsO_4 \cdot H_2O$ [32] and $LiBeAsO_4 \cdot H_2O$ [33]. Because there is no change in the observed space group these framework structures are relatively easily determined from powder diffraction data, and, more importantly, they are relatively easily recognized from their powder patterns.

Powder structure determination becomes more complicated when changes in both cell dimension and symmetry occur and in these cases similar arrangements of the framework atoms can go unrecognized. An example is the case of zeolites with the gismondine-type framework arrangement (GIS). This framework is extremely flexible and changes in both the framework and non-framework atoms cause structural changes. Several minerals, with different compositions, but with the GIS framework arrangement with topological symmetry $I4_1/amd$, have been refined using single crystal data in different space groups (such as $I112/b$, $Fddd$, $P2_1c$, $P2_1$, $P2_12_12_1$, $Pnma$, $I2$, $I\bar{4}$, $Pmn2_1$ and $P2_1/a$) [34]. The framework deformation in dehydrated gismondine from Montalto di Castro, Italy ($Ca_{3.91}Al_{7.77}Si_{8.22}O_{32} \cdot 17.57\,H_2O$) was determined by single crystal X-ray diffraction (Fig. 3; [36]). On dehydration the symmetry of this material changes from $P2_1/c$ to $P2_12_12_1$ with a doubling of the unit cell volume combined with an observed shrinkage of the (doubled) unit cell volume by 17%. These subtle symmetry changes, that are easily observed using single crystal data, are much more difficult to observe from powder data. It is much more difficult to recognize from powder data that two framework connectivities are identical when there are significant changes in the space group and cell dimensions.

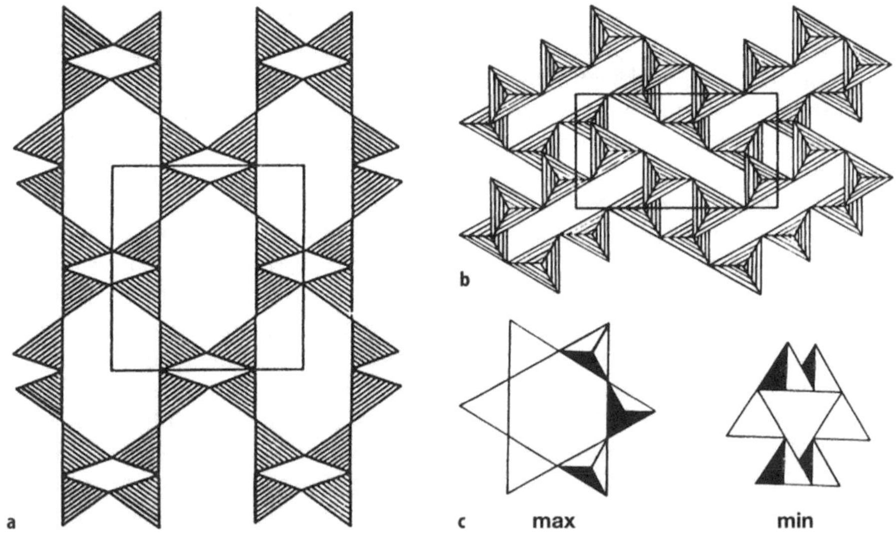

Fig. 2a–c. The ABW framework. **a** The maximum volume unit cell projected down [001]: $a=8.96$, $b=9.50$, $c=5.49$ Å. Approximately the same unit cell is observed in anhydrous Cs-ABW. **b** The minimum volume unit cell projected down [001]: $a=9.50$, $b=5.49$, $c=4.48$ Å. Approximately the same unit cell is observed in anhydrous Li-ABW. **c** Six-ring in the maximum (max) and minimum (min) volume unit cell seen along [100] with [001] vertical. Reproduced by permission of Elsevier Science from [29]

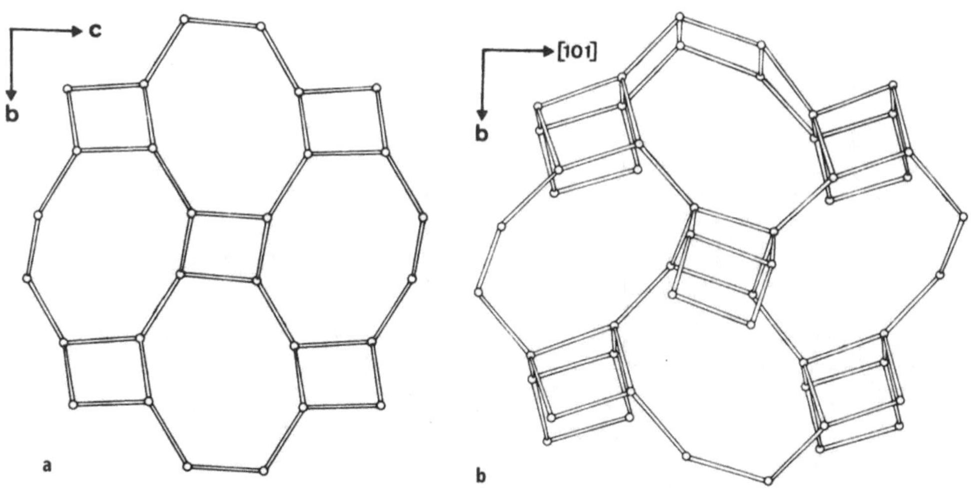

Fig. 3a, b. Framework deformation in gismondine: **a** non-dehydrated (space group: $P2_1/c$), **b** after dehydration in vacuum at room temperature for 24 h (space group: $P2_12_12_1$). Reproduced by permission of Elsevier Science from [36]

For zeolites with the RHO topology [32, 39] the cubic unit cell dimension varies from $a = 13.100$ Å in the dehydrated beryllophosphate mineral pahasapaite [44] to $a = 15.098$ Å in dehydrated deuterium exchanged rho [49]. If a is larger than 14.95 Å the centrosymmetric space group *Im3m* is observed and if a is smaller than this the acentric space groups $I\bar{4}3m$ or *I23* are observed. In the centrosymmetric form the double 8-rings are essentially circular, while in the acentric forms the 8-rings are elliptical (Fig. 4). The ellipticity parameter (EL) is a measure of the difference between the major and minor axes of the elliptical 8-ring [42] and is a function of both the cation type present and the degree of hydration. A regression analysis of the acentric aluminosilicate framework structures gave EL = 13.265 – 0.798 x (unit cell dimension) [48]. A similar linear variation of the EL parameter against the cubic unit cell dimension is observed for the beryllophosphates but with an offset from the trend observed for the aluminosilicates due to the differences in radii of beryllium and phosphorus as compared to those of aluminum and silicon [46]. All the structural details were obtained from powder diffraction data, except for those for the mineral pahasapaite. The datum for pahasapaite, determined from a single crystal structure determination, lays on the regression line determined from the powder data and so strengthens the significance of these data.

Extensive studies on zeolites with NAT [25, 50] or SOD [54] topology have shown that, in order to accommodate different sizes of non-framework species, the frameworks collapse by tilting, shearing and/or deformation of the TO_4 tetrahedra. In the NAT frameworks this flexibility leads to a variety of space groups (e.g., $I\bar{4}2d$, *Fdd2, F1d1, Fd11, C112* and *F2*), and most of these structures have been determined by single crystal diffraction. In contrast, nearly all SOD frameworks exhibit the same ($P\bar{4}3n$) space group symmetry and nearly all these structures have been determined by powder diffraction. Therefore, it seems reasonable to state that, when only powder samples are available, a powder pattern refinement is more successful when the choice of the possible space groups is limited and the number of refinable parameters is small.

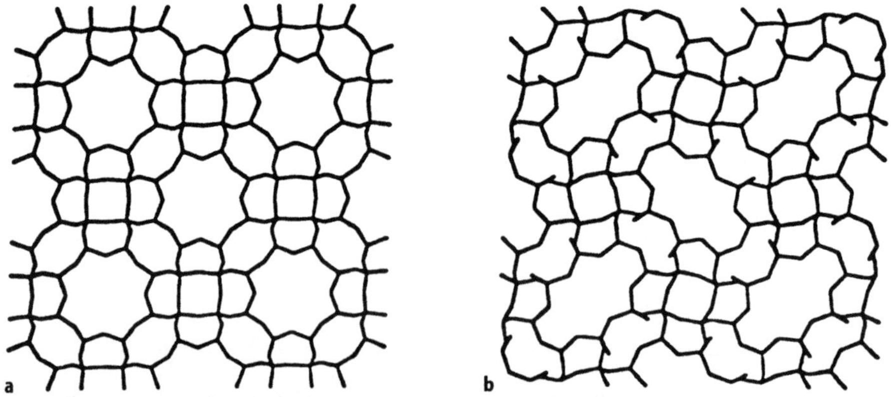

Fig. 4a, b. The centric (a) and acentric (b) form of the RHO framework projected down [001]. Reproduced by permission of Elsevier Science from [45]

Subtle symmetry changes are frequently deduced from diffraction data. Lowering of symmetry usually increases the number of variable parameters and also increases the number of reflections which, especially with powder data, can complicate the analysis of the symmetry changes enormously because of overlapping reflections. A successful explanation of such a subtle symmetry change, caused by shearing of TO_4 layers, has been given from the single crystal X-ray diffraction data of high-silica zeolite ZSM-5 (MFI). The structure of as-synthesized ZSM-5, containing the tetrapropylammonium (TPA) ion, was described using the orthorhombic space group *Pnma* [71]. The empty, calcined framework, H-ZSM-5, shows a reversible displacive phase transition at about 340 K. The precise transition temperature is dependent on the number and type of atoms substituting for the framework silicon atoms [72, 73]. H-ZSM-5 exhibits monoclinic symmetry below and orthorhombic symmetry above this transition temperature. The high-temperature H-ZSM-5 phase is a single crystal with the same orthorhombic *Pnma* symmetry and geometry as the as-synthesized ZSM-5 crystal (containing TPA) (Fig. 5a; [74]) and it is concluded that the template does not deform the framework significantly. Upon cooling, the empty orthorhombic *Pnma* crystal changes into an aggregate of twin domains with monoclinic $P2_1/n11$ symmetry.

Rotation photographs from a H-ZSM-5 crystal at different temperatures (Fig. 6) illustrate this phase transition. At 295 K splitting of the reflection spots is observed. From these photographs and the framework topology it can be concluded that the twin formation can be ascribed to a mutual shift (a shear) of successive (010) pentasil layers along the +c or −c axis with equal probability

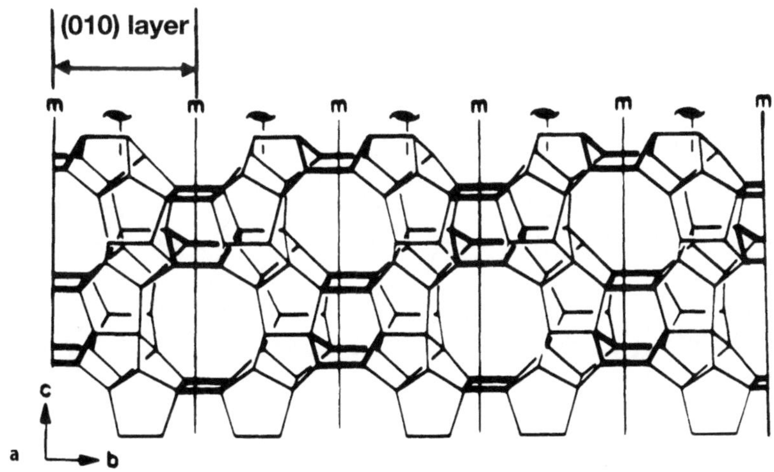

Fig. 5. **a** (100) pentasil layer in H-ZSM-5 with orthorhombic *Pnma* symmetry. **b** (100) Pentasil layer in monoclinic H-ZSM-5 at room temperature. Random (exaggerated) shift of (010) layers along +c and −c, leading to a twinned crystal with $P2_1/n11$ symmetry. The size of the twin domains in the actual crystal is at least about 50 unit cells (~1000 Å). **c** Monoclinic H-ZSM-5 after application of mechanical stress. A perfect monoclinic single crystal is shown. **d** (100) Pentasil layer showing the strictly alternating shift of successive (010) layers along c, leading to orthorhombic $P2_12_12_1$ symmetry

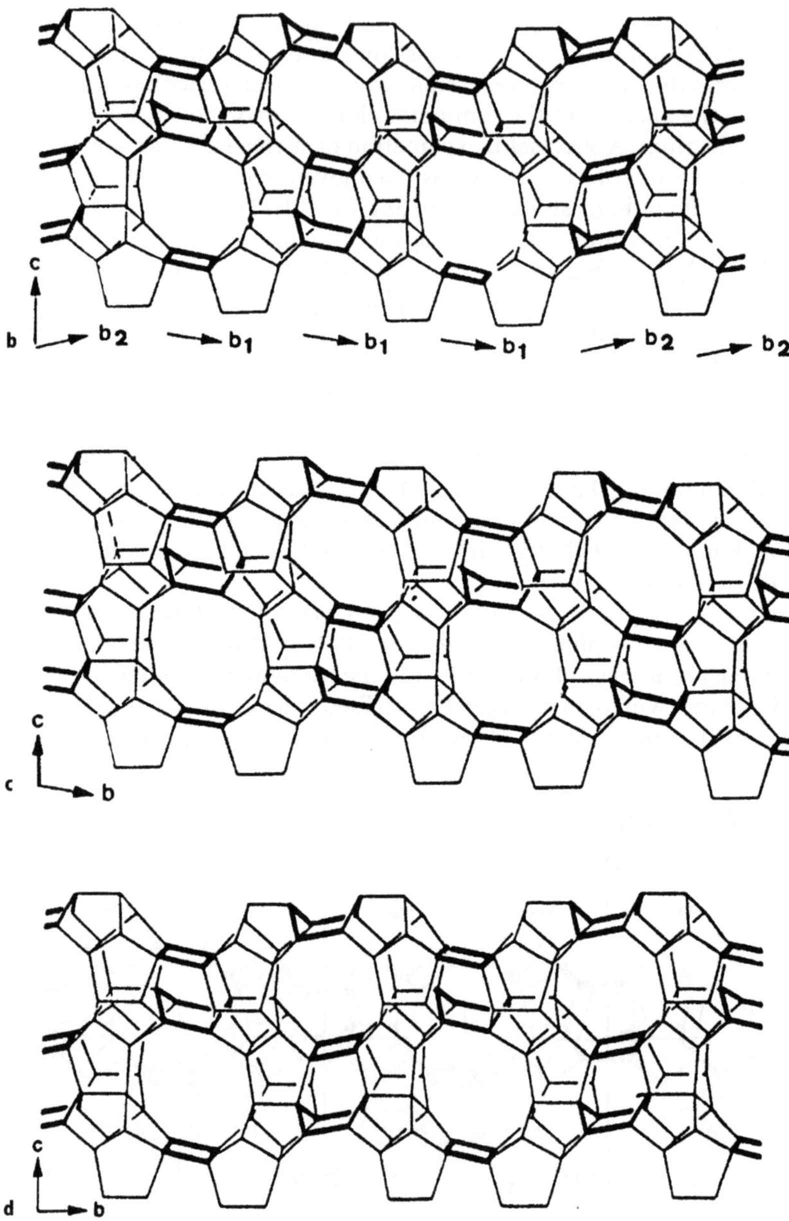

Fig. 5 (continued)

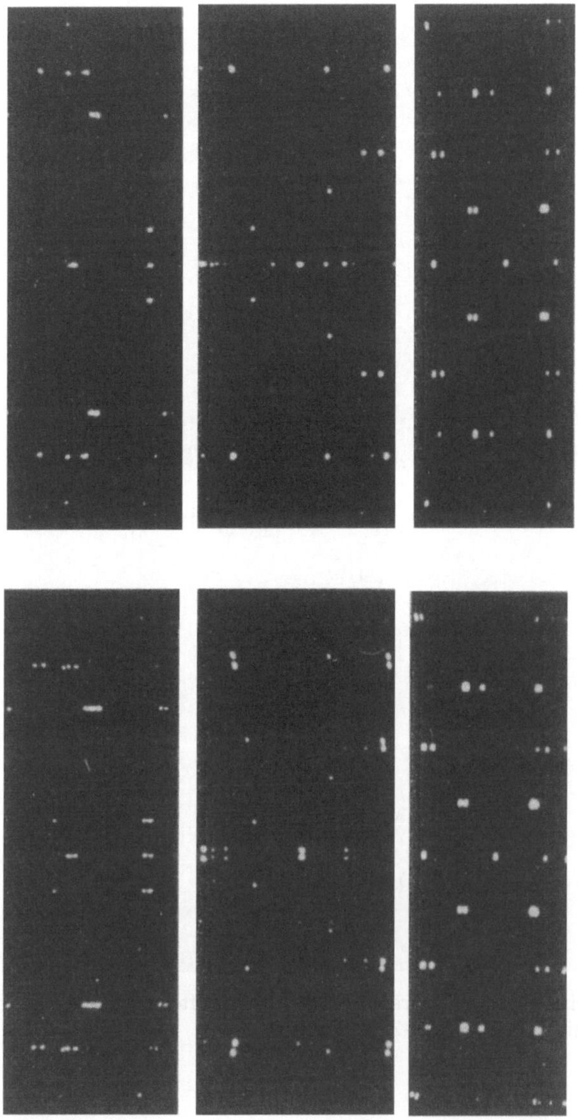

Fig. 6. Details of rotation photographs of a H-ZSM-5 crystal around [100], [010] and [001] (*left to right*) at 400 K (*top*) and 295 K (*bottom*). The rotation axis runs vertical in the plane of the paper

(Fig. 5b; [27]). H-ZSM-5 appears to be ferroelastic: application of an appropriate uniaxial mechanical stress during the orthorhombic/monoclinic transition changes the population of the monoclinic twin domains and a monoclinic (nearly) single crystal can be produced (Fig. 5c; [76]). From Fig. 7 it can be seen that the ratio of the intensities in the 0kl doublets change drastically upon application of a uniaxial mechanical stress. The volume fraction of one of the twin

a

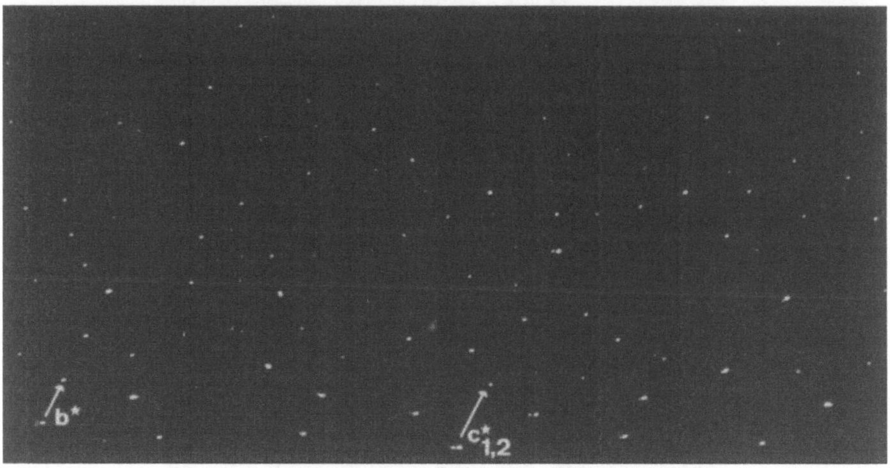

b

Fig. 7 a, b. 0kl-Weissenberg photographs before (**a**) and after (**b**) application of an appropriate uniaxial mechanical stress on a H-ZSM-5 crystal

domains changes from 0.5 to 0.06 after application of this mechanical stress to the crystal used for structure determination [77]. At room temperature, the monoclinic/orthorhombic symmetry change can be reversibly induced by sorption/desorption of various organic molecules (e.g., *p*-xylene, *p*-dichlorobenzene, *p*-nitroaniline and naphthalene [78]). The sorbate loaded and sorbate free H-ZSM-5 shows orthorhombic and monoclinic symmetry, respectively. At low sorbate loading, when there are (sufficient) sorbate molecules in the straight channels only, H-ZSM-5 exhibits the orthorhombic space group *Pnma* [78, 80, 86]. High sorbate loadings, when there are additional sorbate molecules in the sinusoidal channels, bring about yet another symmetry change. The shift of adjacent (010) pentasil layers along c now strictly alternates and the H-ZSM-5 framework transforms to orthorhombic symmetry with space group $P2_12_12_1$ (Fig. 5 d; [78, 79, 82, 84]). All these symmetry changes were studied using single crystal data. It would be nearly impossible to determine these changes using powder diffraction data. Moreover, since the ratio between the number of observations and the number of unknowns is dangerously smaller than one, it would be even more difficult to refine the data.

Zeolite A (LTA) is an example of a non-collapsible framework. The flexibility of this framework has been effectively summarized by Baur (Fig. 8). Single crystal and powder structural data from 108 determinations were extracted from ZeoBase [87]. In one extreme configuration, the T–O1–T angle is almost 180° with the corresponding T–O2–T angle of almost 128°, and in the opposite configuration the angles are reversed. The size and shape of the 8-ring is therefore almost the same in the two extreme configurations but rotated 45° with respect to each other (Fig. 9). For a circular ring opening both T–O–T angles should be close to 155°; the framework is very flexible at these intermediate

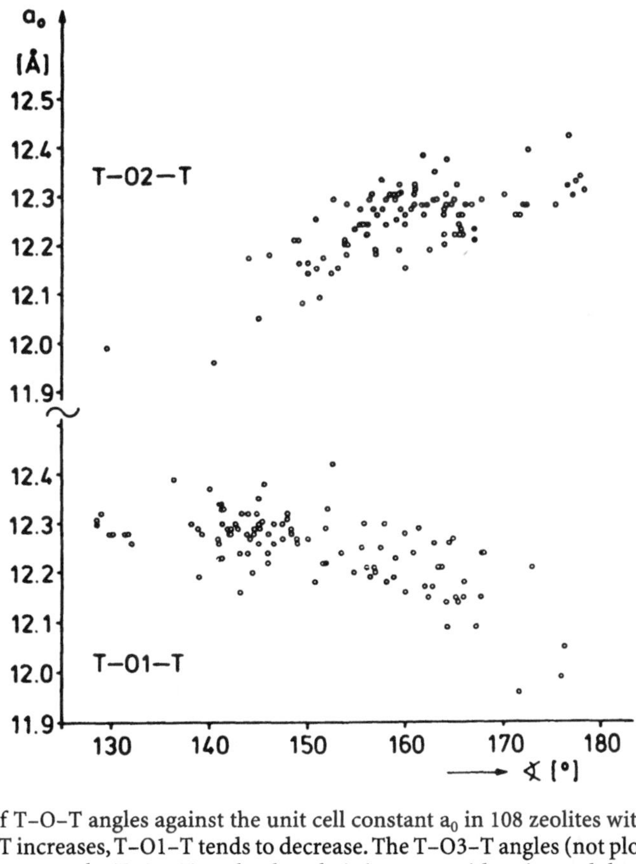

Fig. 8. Plots of T–O–T angles against the unit cell constant a_0 in 108 zeolites with LTA topology. As T–O2–T increases, T–O1–T tends to decrease. The T–O3–T angles (not plotted) increase in the same sense as the T–O2–T angles, but their increase with a_0 is much less. Reproduced by permission of Academic Press from [28]

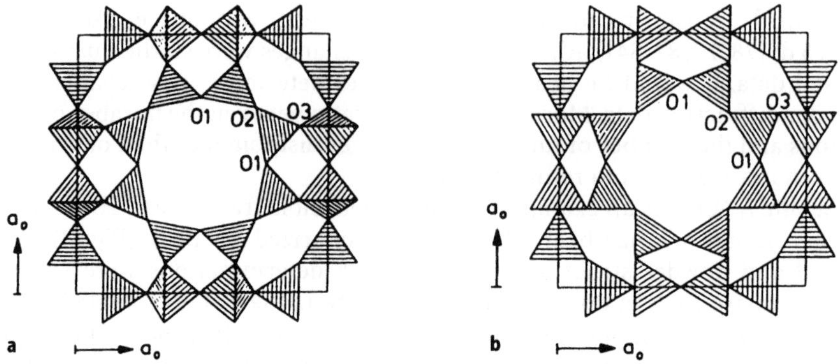

Fig. 9 a, b. The two extremes of possible distortions in LTA: **a** Dehydrated K-exchanged zeolite A; $a_0 = 12.31$ Å; T–O1–T = 128.5°, T–O2–T = 178.4°, T–O3–T = 153.7°. **b** Dehydrated Li-exchanged zeolite A; $a_0 = 11.96$ Å; T–O1–T = 171.6°, T–O2–T = 140.4°, T–O3–T = 133.4°. Reproduced by permission of Academic Press from [28]

angles. Furthermore the effective size of the pore openings in Zeolite A can be modified by the appropriate choice of exchangeable cations which partially block the pore windows. In such a way pore cross sections of 3 Å (K^+ exchanged form), 4 Å (Na^+ exchanged form), or 5 Å (Ca^{2+}/Na^+ form) can be produced. The symmetry changes from *Pm3m* to approximately *Fm3c* with a corresponding eight fold increase in unit cell volume. In most cases only a few very weak reflections are available to support refinement in *Fm3c*. Therefore, many single crystal structure refinements were carried out in the higher symmetry space group *Pm3m* [88]. However, because the higher symmetry space group constrains the framework atoms on more special positions, many of the calculated interatomic distances are in error. A successful refinement in *Fm3c* has been reported for a fluoride containing $GaPO_4$-LTA using powder data [91] illustrating the potential of current powder diffraction methods.

5 Disorder of Non-Framework Species

Another crystallographic problem inherent to zeolite structure analysis is the localization of non-framework species. The often high symmetry of the framework is rarely obeyed by the guests such as templates, adsorbed molecules or cations leading to partial occupancies, disorder and pseudo-symmetry. In nearly all zeolite structures presently studied, the occluded material is disordered. When the point group symmetry of the site where the guest molecule resides is much higher than the symmetry of the guest molecule itself, the induced disorder is many fold and an accurate determination of the position and geometry of the extra-framework molecules becomes very difficult. If, in addition, the occluded material partially occupies two or more positions not related by a symmetry operation of the space group, the electron density of the atoms is spread over many sites within the zeolite and the localization of the material becomes nearly impossible.

Very highly disordered template molecules can sometimes be successfully modeled as molecules with spherical electron density such as the guest molecules quinuclidine in $AlPO_4$-16 (AST; [92]) and 1-aminoadamantane in dodecasil-1H (DOH; [93]). A novel chiral zincophosphate, (CZP; [94,95]), very probably contains a (disordered) infinite helix built up of sodium cations and water molecules. The structural role of the non-framework species is important here because the framework, which is stable under ambient conditions, irreversibly collapses to a condensed structure on dehydration.

In some cases fully ordered positions of the template molecules have been observed and the structure analysis provided valuable information as to the host-guest interaction. For example, the triple helix of water molecules in VPI-5 (VFI; [19]) probably plays an important role as structure directing agent and exhibits the same symmetry as the framework (Fig. 10). In this structure there is no disorder and a very accurate description of the structure was obtained, even though the organic molecule (di-n-propylamine), present in the synthesis mixture, could not be found in the refined structure. This observation raises an

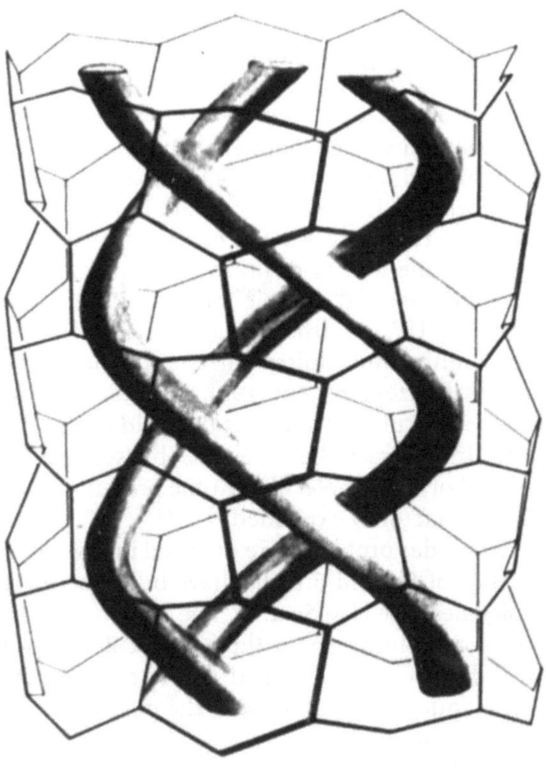

Fig. 10. The triple water helix, represented by the three grey tubes, within the 18-ring channel in VPI-5. Reproduced by permission of Elsevier Science from [19]

important question, that cannot be answered here, as to the role of the di-n-propylamine in the synthesis of VPI-5.

Another example of template ordering was observed for cobalticinium nonasil (NON; [96]). The point group symmetry of the cobalticinium template cation, $Co(C_5H_5)^{2+}$, is *2/m* and its center of symmetry coincides with the center of symmetry of the large cage (point group $P\bar{1}$) and the template ion is perfectly ordered. In the final example of template ordering the framework symmetry of $AlPO_4$-34 (CHA; [97]) is very low ($P\bar{1}$) and the morpholinium template resides in a fixed general position. The template cations tetraethylammonium in $AlPO_4$-18 (AEI; [98]), tetrapropylammonium in SAPO-40 (AFR; [17, 18]), 18-crown-6-Na^+ in EMC-2 (EMT; [99]) and DABCO in CoGaPO-5 (CGF; [100]), are only two-fold disordered because of the small difference in symmetry between the templates and the frameworks at the position of the templates. The results for NON, CHA and CGF were obtained from single crystal data and those for AEI, AFR and EMT from high-resolution powder data. Thus with either technique, accurate information on the localization of the template can be obtained as long as their positions are completely ordered or only slightly disordered.

The same order/disorder problems arise when (organic) molecules are adsorbed into the zeolites and only a few examples of a successful localization of the adsorbate within the framework have been reported. The study on the structure of disordered *m*-xylene sorbed on barium exchanged X (FAU; [101]) at different loadings shows that with careful attention to detail excellent results can be obtained from powder data. The study revealed that when the loading increases, different molecular orientations are adopted by the *m*-xylene molecules in order to maximize methyl–methyl distances and minimize the intermolecular repulsion. Other examples which illustrate the potential of accurate powder diffraction are the studies of various organic molecules adsorbed into type Y (FAU) zeolites. The sites of aniline and *m*-dinitrobenzene, simultaneously adsorbed in NaY using selective deuterated organic molecules, were studied by neutron powder diffraction [102]. UV spectroscopy gives evidence of a charge transfer interaction between aniline and dinitrobenzene. In the rare earth exchanged Na,YbY/1,3,5- trimethylbenzene system [103], the mesitylene molecules occupy two distinct sites. The molecules on site I are two-fold disordered, while site II is only singly occupied. In contrast, the other Na,YbY/sorbate systems [104, 105] show highly disordered organic molecules. At the present time these differences in results cannot be satisfactorily explained.

No serious order/disorder problems are involved in several H-ZSM-5/sorbate systems. Single crystals of H-ZSM-5 (MFI) have been successfully loaded with several organic molecules (see also Sect. 4). The structure of a single crystal of low-loaded H-ZSM-5, containing about three molecules of *p*-dichlorobenzene (pdcb) per unit cell, has been determined in the orthorhombic space group *Pnma* [83]. The sorbed pdcb molecules prefer the position at the intersection of channels (Fig. 11a, b). Although the symmetry of the pdcb molecule is compatible with the site symmetry of the framework it turns out that the molecular mirror plane perpendicular to the Cl–Cl axis does not coincide with the crystallographic mirror plane and a 2-fold positional disorder around the mirror plane

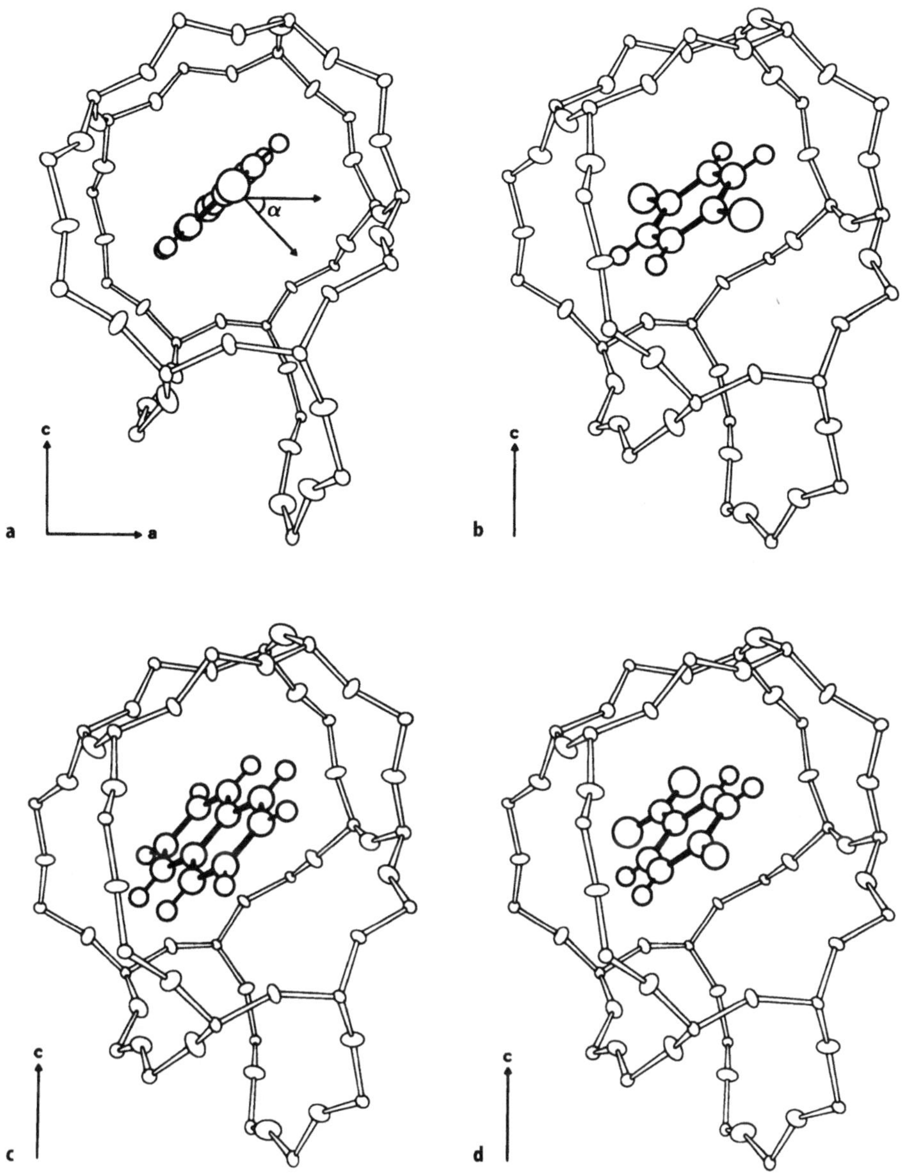

Fig. 11 a–d. ORTEP drawings [106] of the position and orientation of adsorbed molecules at the intersection of channels in low-loaded H-ZSM-5. Open bonds connect framework atoms and solid bonds connect atoms in adsorbed molecules. **a** *p*-dichlorobenzene molecules in H-ZSM–5/2.6 *p*-dichlorobenzene, viewed down the straight channel axis. **b** as in **a** but viewed down an axis inclined 20° with the straight channel axis. **c** naphthalene molecules in H-ZSM-5/3.7 naphthalene, viewed as in **b**. **d** *p*-nitroaniline molecules in H-ZSM-5/4.0 *p*-nitroaniline, viewed as in **b**

Table 2. Orientation of adsorbates at the intersection of channels in H-ZSM-5

Code[a]	PDCB1	NAPH	PNAN[b]	PXYL	PDCB2
x	0.4860	0.4860	0.4858	0.4894	0.4824
y	0.2400	0.2366	0.2332	0.2379	0.2439
z	−0.0188	−0.0352	−0.0260	−0.0180	−0.0175
α[c]	47.1	40.5	44.3	−31.1	−26.2

[a] Codes are as follows:
PDCB1: H-ZSM-5 containing 2.6 molecules *p*-dichlorobenzene/u.c.
NAPH: H-ZSM-5 containing 3.7 molecules naphthalene/u.c.
PNAN: H-ZSM-5 containing 4.0 molecules *p*-nitroaniline/u.c.
PXYL: H-ZSM-5 containing 8.0 molecules *p*-xylene/u.c.
PDCB2: H-ZSM-5 containing 8.0 molecules *p*-dichlorobenzene/u.c.
[b] Molecular center calculated disregarding the oxygen atoms.
[c] The angle α is defined in the text and illustrated in Fig. 11a.

occurs. The location and rotational orientation of the sorbate at the intersection can, in a first approximation, be described by the fractional coordinates (x, y, z) of its molecular center and the angle α between the positive a axis and the vector normal to the aromatic ring plane (Fig. 11a; Table 2).

In single crystals of H-ZSM-5 loaded with four molecules of naphthalene [81] or four molecules of *p*-nitroaniline [86] per unit cell, both exhibiting orthorhombic *Pnma* symmetry, the organic molecules at the intersection are in an analogous orientation as in the low-loaded H-ZSM-5/pdcb system (Fig. 11b, d; Table 2). In H-ZSM-5, fully loaded with eight molecules *p*-xylene per unit cell, the adsorbate has been found to be ordered in the orthorhombic space group $P2_12_12_1$, allowing its packing determination (Fig. 12a; [79]). One of the *p*-xylene molecules lies at the intersection of the straight and sinusoidal channels with its long molecular axis nearly parallel to (100) and deviating about eight degrees from the straight channel axis.

The second *p*-xylene molecule is in the sinusoidal channel. Its long molecular axis is practically parallel to (010) and deviates almost six degrees from [100]. The structural aspects of H-ZSM-5 loaded with eight *p*-dichlorobenzene (pdcb) molecules per unit cell [84] are in all details comparable to those in the high-loaded H-ZSM-5/8 *p*-xylene system (compare Figs. 12a and 12b).

The phase transition from *Pnma* to $P2_12_12_1$ can be connected to a sudden increase in ordering of the sorbed phase with increasing coverage [79, 107, 108]. This commensurate crystallization of molecules within the H-ZSM-5 framework is assumed to be stabilized by establishing contacts at the channel intersection between adjacent molecules (See Fig. 12; [109]). However, the methyl(H)–aromatic ring interactions in H-ZSM-5/8 *p*-xylene are replaced by Cl-ring (C, H) interactions in H-ZSM-5/8 pdcb, which are substantially weaker. The importance of these interactions in stabilizing the guest structure within the zeolite host framework might therefore need reconsideration. The ring-(C, H)–framework (O) contacts, which are the same in both structures, might be important in stabilizing the actually observed packing arrangement.

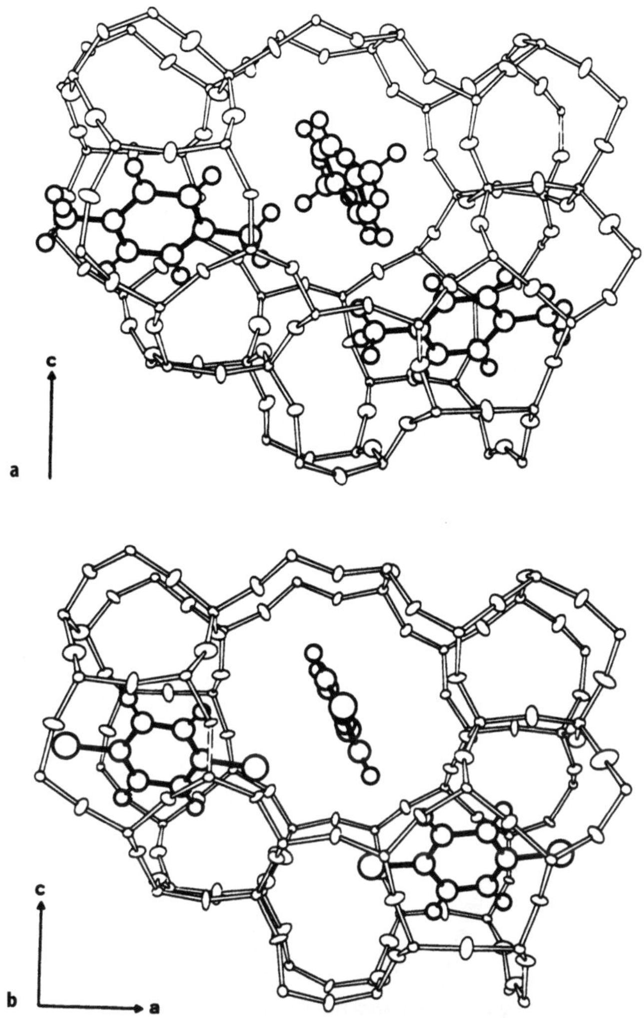

Fig. 12a, b. Position and orientation of adsorbed molecules in high-loaded H-ZSM-5. **a** *p*-xylene molecules at the intersection of channels and in the sinusoidal channel in the high-loaded H-ZSM-5/8 *p*-xylene system. The angle between the view direction and the straight channel axis is 15°. **b** As in **a** with *p*-xylene replaced by *p*-dichlorobenzene and viewed down the straight channel axis

The longest dimension of the adsorbed molecule, being similar to the length of the straight channel axis, very probably determines the commensurate crystallization by a flexible response of the channel pores in the zeolite host framework. The rotational orientation of pdcb in the low-loaded system differs 73.3° (=47.1+26.2) and 78.2° (=47.1+31.1) from the rotational orientation of the pdcb and *p*-xylene molecules trapped at the intersection in the high-loaded H-ZSM-5/sorbate systems (see Table 2). These two rotational orientations correspond to the directions of the two maximal pore dimensions observed in the clover-like window in the empty H-ZSM-5 framework at 350 K [74].

From Table 2 it can be seen that the molecular center (disregarding the oxygen atoms) is approximately the same "off-intersection center" position in all structures. The same type of disorder is observed in all low-loaded systems. Refinements of the low-loaded systems show that the starting orientation of the sorbate molecule may be rather far away from its minimized orientation as long as its geometry is reasonable and its molecular center is not too far away from the minimized value. The examples illustrate that sorbed molecules can be located within the zeolite framework by single crystal X-ray diffraction methods when the symmetry of the adsorption site is compatible with the symmetry of the sorbed molecule and also in some cases when these symmetries do not coincide and disorder occurs.

Cations often occupy special positions in the structure and their (assumed) spherical symmetry often coincides with the local framework symmetry. The positions of these cations have been accurately determined in many zeolites (particularly in the zeolites with either the FAU, LTA, NAT, RHO or SOD topology). One example is the single crystal structure of zeolite A (LTA) exchanged with Pb^{2+} at pH = 6.0 and evacuated at 300 K [89] which contains, in each sodalite cage, a distorted Pb_4O_4 cube from which one Pb^{2+} has been removed. The four oxygen atoms that were connected to the missing Pb^{2+} ion, are now coordinated to four additional Pb^{2+} ions thus stabilizing the $[Pb_7O(OH)_3]^{9+}$ ion (Fig. 13). The structure determination illustrates the ability of zeolites to capture solute structures.

It is not possible to cover even a small number of the papers describing many different zeolite/cation systems. The reader should review a compilation of

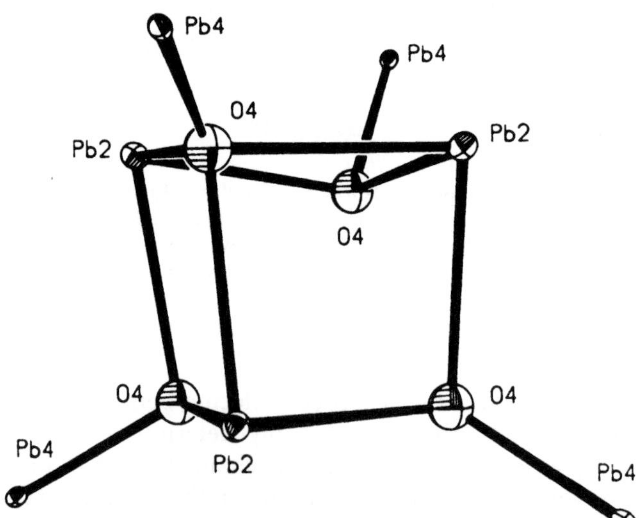

Fig. 13. The $[Pb_7O(OH)_3]^{9+}$ cluster within and extending out of the sodalite unit. Reproduced by permission of Elsevier Science from [89]

extra-framework sites in zeolites [110] published by the Structure Commission of the International Zeolite Association. It is planned to publish an updated issue in 1998, which will include the most recent results. An inexperienced user should be very careful in assigning extra-framework sites to maxima in a difference electron density map because simulations in pseudo-symmetric sodalites [60] and refinement of cation positions in Linde Q (BPH; [111]) have shown that some of the extra-framework electron density observed in a difference map is an artifact caused by the Fourier technique used in the refinement of the framework structure.

6
Faulting Within the Framework

Faulting within the zeolite framework is one of the difficulties encountered in zeolite structure analysis. When faulting occurs repeatedly but in an irregular manner the resulting structure is disordered. In cases where the repeat distances between the faults become small enough diffuse stripes perpendicular to the fault planes are observed in the diffraction pattern. When the fault planes repeat in every unit cell a new framework topology emerges and the resulting diffraction pattern again contains peaks only at Bragg positions. Faulting on a very small and/or irregular frequency will seriously hamper a precise structure determination. Such faulting is usually evident when a powder pattern has broad and sharp peaks. Powder patterns can also have broader and sharper lines when one dimension of the crystal is very thin but this is usually evident from an electron microscopy photograph.

In the classical sense the explanation of faulting and the calculation of the resulting powder pattern should not be called a structure determination, yet it represents one of the more difficult tasks that can be undertaken. Examples are the studies on zeolite beta, FAU-EMT, MFI-MEL, OFF-ERI, RUB-n and the SSZ-n/CIT-n series. It must be emphasized that any time the powder pattern consists of sharp peaks only then faulting is an unsatisfactory explanation for difficulties with a structure determination. An examination of the three proposed end members for zeolite beta [112] shows why the material faults. There are several different sequences of five and six rings around a 12-ring. The only connection between one arrangement and another would be the chosen crystallographic symmetry. But this is an artificial choice, because crystallographic symmetry is a result of the arrangement of the atoms in a structure and not the driving force. The structure of beta has been shown to be an almost random arrangement of sequences and has an almost uniform powder pattern. The accessibility of the pore system is not affected by the degree of faulting nor does it change the diameter of the pore openings. Single crystals of the pure end-members have not been synthesized yet and therefore no single crystal structure analysis of the polymorphs of zeolite beta has been performed [115]. Even the beta mineral analogue, Tschernichite [116], has a powder pattern almost identical to those of the synthesized beta materials.

In contrast, both a wide range of different intermediate phases and unfaulted examples of the end-members can be obtained for other intergrown materials.

Thus, end members of FAU and EMT [117], MFI and MEL, OFF and ERI, some end-members in the RUB-n series (the zincosilicates RUB-17 (RSN; [120]) and VPI-7 (VSV; [121, 122]), the beryllosilicate lovdarite (LOV; [123]), the decasil RUB-3 (RTE; [124]), the borosilicate RUB-13 (RTH; [125])), and one end-member in the SSZ/CIT series (the borosilicate CIT-1 (CON; [126, 127])) can all be produced by a proper choice of synthesis conditions and templates. Blocking of the pores can occur in some of the faulted intergrowths thereby drastically changing the properties of the materials. Only the syntheses of zeolite X (FAU) and ZSM-5 (MFI) currently give single crystals suitable for an accurate X-ray structure analysis [71, 128].

7 Isomorphous Replacement of Framework Atoms

Isomorphous replacement of framework atoms has been studied in several zeolites. The problem with this type of diffraction experiments is that the difference in scattering power of a tetrahedral site with and without the substitution is often small. For example, replacement of 10% of the aluminum atoms at one site by manganese ($(Al_9Mn)P_{10}O_{40}$; AEL; [129]) is equivalent of looking for one extra electron on a site where there is already an atom. The detection of partial replacement of phosphorous by silicon, such as in SAPO-43 (GIS; [35]), is even more difficult.

The incorporation into the framework of tetrahedral atoms with a radius different from the radius of the substituted tetrahedral atom can lead to a change in unit cell volume and X-ray diffraction can be used to follow these changes. Cell expansion or contraction does not conclusively establish the actual incorporation of tetrahedral atoms into the framework, since extra framework species may also change the cell volume. The actual distribution of the incorporated tetrahedral atoms on the framework sites has been studied using diffraction data by examining both the refined T–O distances (and sometimes the O–T–O angles) and the refined population parameters. From statistical studies on the T–O distances it was inferred that in substituted Nu-1-type frameworks (RUT; [130]) the boron atoms were uniformly distributed over all the tetrahedral sites. This conclusion may be in error because both the accepted value for a Si–O distance was used and it was assumed that the T–O–T angle would be unchanged. A recent paper on a material with the same topology (RUB-10; [131]) now indicates that the boron atoms may be ordered but the space group has been changed to $P2_1/a$.

In several of the cited papers the population parameters (or site occupancies) of the framework atoms have been refined. The population parameter is frequently used as a parameter which mimics the success/failure of isomorphous substitution. However, it is dangerous to draw too many conclusions from changes in the refined population parameters because, like the (anisotropic) displacement parameters, they act as a waste-basket for all systematic and non-systematic errors occurring in a refinement. For example, missing tetrahedral atoms influence the population parameter in a non-systematic way and even the actual scattering factors used in the refinement affect the site occupancies. It has also been shown that in dodecasil-1H (DOH; [93]) when the oxygen atoms are disordered, the population factors of the tetrahedral atoms appear smaller than 1.0 and this was shown to be an artifact of the least squares algorithm used. Other examples include the single crystal determination of the structure of ZAPO-M1 (ZON; [132]), where both the distribution of the T–O bond lengths and the population parameters were consistent with a non-uniform distribution of the zinc atoms. It has also been claimed that from very precise single crystal diffraction studies it was observed that the substitution of silicon for phosphorus in SAPO-31 (ATO; [133]) and of cobalt for (exclusively) aluminum in both CoAPO-5 (AFI; [134]) and CoSAPO-34 (CHA; [135]) had taken place. In many other structure reports the opposite conclusion was reached, namely that replacement of framework atoms could not be established from the analysis of X-ray diffraction data alone.

8 Crystal Size Limitations

The possibility of using single crystal data rather than powder X-ray diffraction became much more feasible with the development of area detectors for use with high-intensity synchrotron X-ray sources. A 30×30×30 μm crystal of (Mg,Al)PO_4-STA-1 (SAO; [136]) was successfully used for structure determination at the ESRF in Grenoble using a diffractometer equipped with a CCD detector. Unfortunately, no low angle data were collected which probably accounted for why the template could not be localized. Data from crystals of this size have been previously collected using rotating anode generators [34] and even sealed tube data from crystals of about the same volume [137] have been measured. Unfortunately, the use of these data was restricted to only determining the framework structure. In the structure analysis of a 35×20×15 μm crystal of $AlPO_4$-34 (CHA; [97]) synchrotron diffraction data were collected at the SRS at Daresbury and used to determine the location of the morpholinium cation. Theoretically the coherence length of about 0.1 μm determines the lower size limit for single crystal diffraction [138] and the physical ability to handle such small crystals may finally prove to be the limiting factor in their use.

As the size of the crystal that can be used decreases it can be expected that the use of single crystal diffraction facilities at synchrotron sources by the zeolite community will increase enormously in the coming years. However, many new microporous materials are synthesized with crystals that are only 1 μm in size – still too small to be utilized for single crystal work – and only can be studied by powder techniques.

9 Conclusions

As hopefully explained here, the problems with attempting structure determinations of unknown materials requires that an accurate starting model be obtained and all of the possible problems that could cause subtle framework changes be evaluated. This requires that the unit cell dimensions and space group be correctly determined. Molecular modeling techniques have started to be applied to determining starting models. They are currently restricted by being difficult to use to compose a framework arrangement within a defined unit cell and space group. Other techniques are focusing on using the power of current computers to try to determine a framework arrangement by starting with random arrangements of atoms and hoping that the program can refine the starting models to the correct topology.

While all these techniques will continue to improve over the next few years, the quality and accuracy of the final crystal structure will depend on how good the data are and how well the unit cell and space group have been determined. While not covered in this paper, it should be also realized that input from many other characterization techniques should be used as an aid in the determination of an unknown structure, that the final solution should be in agreement with data from all the different techniques employed and that the solution should make chemical sense. When the structure has been correctly determined, it is possible to use the information to explain framework distortions, cation and sorbate locations, isomorphous substitutions and other subtle variations in the behavior of the structure. But, the use of the structure of a material requires a correct structure determination and obtaining that solution is both a science and an art and not a trivial exercise.

References

1. McCusker LB (1991) Acta Cryst A47:297
2. Baerlocher Ch, McCusker LB (1994) Practical aspects of powder diffraction data analysis. In: Jansen JC, Stöcker M, Karge HG, Weitkamp J (eds) Advanced zeolite science and applications. Stud Surf Sci Catal Vol 85. Elsevier Science, Amsterdam, p 391
3. Visser JW (1993) A fully automatic program for finding the unit cell from poweder data. Version 15. Method described in (1969) J Appl Cryst 2:89
4. Werner PE (1990) Treor 90. A trial and error program for indexing of unknown powder patterns. Department of Structural Chemistry, Arrhenius Laboratory, University of Stockholm
5. Jansen J, Peschar R, Schenk H (1992) J Appl Cryst 25: 237
6. Jansen J, Peschar R, Schenk H (1993) Z Kristallogr 206:33
7. Spengler R, Zimmermann H, Burzlaff H, Jansen J, Peschar R, Schenk H (1994) Acta Cryst B50: 578
8. Estermann MA, McCusker LB, Baerlocher Ch (1992) J Appl Cryst 25:539
9. McCusker LB, GrosseKunstleve RW, Baerlocher Ch, Yoshikawa M, Davis ME (1996) Microporous Mater 6:295
10. Estermann MA, McCusker LB, Baerlocher Ch, Merrouche A, Kessler H (1991) Nature 352: 320
11. Baerlocher Ch, Hepp A, Meier WM (1977) DLS. A Fortran program for the simulation of crystal structures by geometric refinement. Institut für Kristallographie, ETH, Zürich, Switzerland
12. Young RA (1993) The Rietveld Method, Oxford University Press, Oxford
13. Lawton SL, Rohrbaugh WJ (1990) Science 247:1319
14. Bennett JM, Marcus BK (1988) The crystal structures of several metal aluminophosphate molecular sieves. In: Grobet PJ et al. (eds) Innovation in zeolite material science. Elsevier Science, Amsterdam, p 269

15. Harvey G, Baerlocher Ch (1992) Z Kristallogr 201:113
16. Kirchner RM, Bennett JM (1994) Zeolites 14:523
17. Dumont N, Gabelica Z, Derouane EG, McCusker LB (1993) Microporous Mater 1:149
18. McCusker LB, Baerlocher Ch (1996) Microporous Mater 6:51
19. McCusker LB, Baerlocher Ch, Jahn E, Bülow M (1991) Zeolites 11:308
20. Li HX, Davis ME, Higgins JB, Dessau RM (1993) J Chem Soc Chem Commun 1993:403
21. Vogt ETC, Richardson JW Jr (1990) J Solid State Chem 87:469
22. Annen MJ, Young D, Davis ME (1991) J Phys Chem 95: 1380
23. Prasad S, Balakrishnan I (1990) Inorg Chem 29: 4830
24. Richardson JW Jr, Vogt ETC (1992) Zeolites 12: 13
25. Baur WH (1995) Framework mechanics: limits to the collapse of tetrahedral frameworks. In: Rozwadowski M (ed) Proceedings of the 2nd Polish-German zeolite colloquium. Nicholas Copernicus University Press, Torun, p 171
26. Baur WH (1992) Why the open framework of zeolite A does not collapse, while the dense framework of natrolite is collapsible. In: Rozwadowski M (ed) Proceedings of the Polish-German zeolite colloquium. Nicholas Copernicus University Press, Torun, p 11
27. Koningsveld H van, Jansen JC, Bekkum H van (1987) Zeolites 7: 564
28. Baur WH (1992) J Solid State Chem 97: 243
29. Norby P, Fjellvåg H (1992) Zeolites 12: 898
30. Krogh Anderson IG, Krogh Anderson E, Norby P, Colella C, de'Gennaro M (1991) Zeolites 11: 149
31. Newsam JM (1988) J Phys Chem 92: 445
32. Gier TE, Stucky GD (1991) Nature 349: 508
33. Harrison WT, Gier TE, Stucky GD (1995) Acta Cryst C51: 181
34. Pluth JJ, Smith JV, Bennett JM (1989) J Am Chem Soc 111: 1692
35. Helliwell M, Kaučič V, Cheetham GMT, Harding MM, Kariuki BM, Rizkallah PJ (1993) Acta Cryst B49: 413
36. Vezallini G, Quartieri S, Alberti A (1993) Zeolites 13: 34
37. Hansen S, Håkansson U, Fälth L (1990) Acta Cryst C46: 1361
38. Hansen S, Håkansson U, Landa-Canovas AR, Fälth L (1993) Zeolites 13: 276
39. Baur WH, Fischer RX, Shannon RD (1988) Relations and correlations in zeolite RHO and computer simulations of its crystal structure. In: Grobet PJ et al. (eds) Innovation in zeolite materials science. Elsevier Science, Amsterdam, p 281
40. Fischer RX, Baur WH, Shannon RD, Staley RH, Abrams L, Vega AJ, Jorgensen JD (1988) Acta Cryst B44: 321
41. Baur WH, Bieniok A, Shannon RD, Prince E (1989) Z Kristallogr 187: 253
42. Fischer RX, Baur WH, Shannon RD, Parise JB, Faber J, Prince E (1989) Acta Cryst C45: 983
43. Bieniok A, Baur WH (1993) Acta Cryst B49: 817
44. Corbin DR, Abrams L, Jones GA, Harlow RL, Dunn PJ (1991) Zeolites 11: 364
45. Parise JB, Corbin DR, Gier TE, Harlow RL, Abrams L, Dreele RB von (1992) Zeolites 12: 360
46. Parise JB, Corbin DR, Abrams L, Northrup P, Rakovan J, Nenoff TM, Stucky GD (1994) Zeolites 14: 25
47. Parise JB, Corbin DR, Abrams L (1995) Microporous Mater 4: 99
48. Newsam JM, Vaughan DEW, Strohmaier KG (1995) J Phys Chem 99: 9924
49. Baur WH, Fischer RX, Shannon RD, Staley RH, Vega AJ, Abrams L, Corbin DR (1987) Z Kristallogr 179: 281
50. Ståhl K, Hanson J (1994) J Appl Cryst 27: 543
51. Ståhl K, Thomassen R (1994) Zeolites 14: 12
52. Joswig W, Baur WH (1995) N Jb Miner Mh 1995: 26
53. Baur WH, Joswig W (1996) N Jb Miner Mh 1996: 171
54. Richardson JW Jr, Pluth JJ, Smith JV, Dytrych WJ, Bibby DM (1988) J Phys Chem 92: 243
55. Fleet ME (1989) Acta Cryst C45: 843
56. Kempa PB, Engelhardt G, Buhl JCh, Felsche J, Harvey G, Baerlocher Ch (1991) Zeolites 11: 558

57. Nenoff TM, Harrison WTA, Gier TE, Stucky GD (1991) J Am Chem Soc 113: 378
58. Sieger P, Wiebcke M, Felsche J (1991) Acta Cryst C47: 498
59. Depmeier W (1992) Z Kristallogr 199: 75
60. Hu X, Depmeier W (1992) Z Kristallogr 201: 99
61. Depmeier W, Melzer R, Hu X (1993) Acta Cryst B49: 483
62. Harrison WT, Gier TE, Stucky GD (1994) Acta Cryst C50: 471
63. Féron B, Guth JL, Mimouni-Erddalane N (1994) Zeolites 14: 177
64. Brenchley ME, Weller MT (1994) Zeolites 14: 682
65. Mead PJ, Weller MT (1994) Microporous Mater 3: 281
66. Fütterer K, Depmeier W, Altorfer F, Behrens P, Felsche J (1994) Z Kristallogr 209: 517
67. Duke CVA, Hill SJ, Williams CD (1995) Zeolites 15: 413
68. Mead PJ, Weller MT (1995) Zeolites 15: 561
69. Hassan I (1996) Z Kristallogr 211: 228
70. Dann SE, Weller MT (1996) Inorg Chem 35: 555
71. Koningsveld H van, Bekkum H van, Jansen JC (1987) Acta Cryst B43: 127
72. Gabelica Z, Guth JL (1989) Angew Chem 101: 60
73. Lopez A, Soulard M, Guth JL (1990) Zeolites 10: 134
74. Koningsveld H van (1990) Acta Cryst B46: 731
75. Aizu K (1970) Phys Rev B 2: 754
76. Koningsveld H van, Tuinstra F, Jansen JC, Bekkum H van (1989) Zeolites 9: 253
77. Koningsveld H van, Jansen JC, Bekkum H van (1990) Zeolites 10: 235
78. Mentzen BF (1988) J Appl Cryst 21: 266
79. Koningsveld H van, Tuinstra F, Bekkum H van, Jansen JC (1989) Acta Cryst B45: 423
80. Mentzen BF, Sacerdote-Peronnet M, Berar JF, Lefebvre F (1993) Zeolites 13: 485
81. Koningsveld H van, Jansen JC (1996) Microporous Mater 6: 159
82. Mentzen BF, Sacerdote-Peronnet M (1993) Mat Res Bull 28: 1161
83. Koningsveld H van, Jansen JC, Man AJM de (1996) Acta Cryst B52: 131
84. Koningsveld H van, Jansen JC, Bekkum H van (1996) Acta Cryst B52: 140
85. Reck G, Marlow F, Kornatowski J, Hill W, Caro J (1996) J Phys Chem 100: 1698
86. Koningsveld H van, Koegler JH (1997) Microporous Mater 9: 71
87. Baur WH, Fischer RX (1995) ZeoBase, Frankfurt and Mainz
88. Heo NH, Seff K (1992) Zeolites 12: 819
89. Ronay Ch, Seff K (1993) Zeolites 13: 97
90. Jang SB, Kim Y, Seff K (1994) Zeolites 14: 262
91. Simmen A, Patarin J, Baerlocher Ch (1993) Rietveld refinement of F-containing $GaPO_4$–L TA. In: Ballmoos R von, Higgins JB, Treacy MMJ (eds) Proceedings from the Ninth International Zeolite Conference, Montreal 1992. Butterworth-Heinemann, Stoneham MA, p 433
92. Bennett JM, Kirchner RM (1991) Zeolites 11: 502
93. Miehe G, Vogt T, Fuess H, Müller U (1993) Acta Cryst B49: 745
94. Rajiè N, Logar NZ, Kauèiè V (1995) Zeolites 15: 672
95. Harrison WTA, Gier TE, Stucky GD, Broach RW, Bedard RA (1996) Chem Mater 8: 145
96. Goor G van de, Freyhardt CC, Behrens P (1995) Z Anorg Allg Chem 621: 311
97. Harding MM, Kariuki BM (1994) Acta Cryst C50: 852
98. Simmen A, McCusker LB, Baerlocher Ch, Meier WM (1991) Zeolites 11: 654
99. Baerlocher Ch, McCusker LB, Chiapetta R (1994) Microporous Mater 2: 269
100. Chippindale AM, Cowley AR. Paper to be submitted
101. Mellot C, Espinat D, Rebours B, Baerlocher Ch, Fischer P (1994) Catal Lett 27: 159
102. Kirschhock C, Fuess H (1997) Microporous Mater 8: 19
103. Czjzek M, Vogt T, Fuess H (1992) Zeolites 12: 237
104. Czjzek M, Vogt T, Fuess H (1991) Zeolites 11: 832
105. Czjzek M, Fuess H, Vogt T (1991) J Phys Chem 95: 5255
106. Johnson CK (1965) ORTEP, Report ORNL. Oak Ridge Nat Lab, Oak Ridge TN, revised June 1970
107. Reischman PT, Schmitt KD, Olson DH (1988) J Phys Chem 92: 5165

108. Richards RE, Rees LVC (1988) Zeolites 8: 35
109. Thamm H (1987) J Phys Chem 91: 8
110. Mortier WJ (1982) Compilation of extra-framework sites in zolites. Butterworth Scientific, Guildford
111. Andries KJ, Bosmans HJ, Grobet PJ (1991) Zeolites 11: 124
112. Newsam JM, Treacy MMJ, Koetsier WT, Gruyter CB de (1988) Proc Roy Soc Lond A420: 375
113. Treacy MMJ, Newsam JM (1988) Nature 332: 249
114. Higgins JB, LaPierre RB, Schlenker JL, Rohrman AC, Wood JD, Kerr GT, Rohrbaugh WJ (1988) Zeolites 8: 446
115. Marler B, Böhme R, Gies H (1993) Single crystal structure analysis of zeolite beta: the superposition structure. In: Ballmoos R von, Higgins JB, Treacy MMJ (eds) Proceedings from the Ninth International Zeolite Conference, Montreal 1992. Butterworth-Heinemann, Stoneham MA, p 425
116. Boggs RC, Howard DG, Smith JV, Klein GL (1993) Am Mineral 78: 822
117. Delprato F, Delmotte L, Guth JL, Huve L (1990) Zeolites 10: 546
118. Burkett SL, Davis ME (1993) Microporous Mater 1: 265
119. Chatelain T, Patarin J, Soulard M, Guth JL (1995) Zeolites 15: 90
120. Röhrig C, Gies H (1995) Angew Chem Int Ed 34: 63
121. Annen MJ, Davis ME, Higgins JB, Schlenker JL (1991) J Chem Soc Chem Commun 1991: 1175
122. Röhrig C, Gies H, Marler B (1994) Zeolites 14: 498
123. Merlino S (1990) Eur J Mineral 2: 809
124. Marler B, Grünewald-Lüke A, Gies H (1995) Zeolites 15: 388
125. Vortmann S, Marler B, Gies H, Daniels P (1995) Microporous Mater 4: 111
126. Lobo RF, Pan M, Chan I, Li HX, Medrud RC, Zones SI, Crozier PA, Davis ME (1993) Science 262: 1543
127. Lobo RF, Davis ME (1995) J Am Chem Soc 117: 3766
128. Olson DH (1995) Zeolites 15: 439
129. Pluth JJ, Smith JV, Richardson JW Jr (1988) J Phys Chem 92: 2734
130. Bellusi G, Millini R, Carati A, Maddinelli G, Gervasini A (1990) Zeolites 10: 642
131. Gies H, Rius J (1995) Z Kristallogr 210: 475
132. Marler B, Patarin J, Sierra L (1995) Microporous Mater 5: 151
133. Baur WH, Joswig W, Kassner D, Kornatowski J, Finger G (1994) Acta Cryst B50: 290
134. Chao KJ, Sheu SP, Sheu HS (1992) J Chem Soc Faraday Trans 88: 2949
135. Nardin G, Randaccio L, Kauèiè V, Rajiè N (1991) Zeolites 11: 192
136. Noble GW, Wright PA, Lightfoot P, Morris RE, Hudson KJ, Kvick A, Graafsma H (1997) To be published
137. Flanigen EM, Bennett JM, Grose RW, Cohen JP, Patton RL, Kirchner RM (1978) Nature 271: 512
138. Schlenker JL, Peterson BK (1996) J Appl Cryst 29: 178

Structural Information from Neutron Diffraction

A. N. Fitch[1] and H. Jobic[2]

[1] ESRF, BP220, 38043 Grenoble Cedex, France; *e-mail:* fitch@esrf.fr
Department of Chemistry, Keele University, Staffordshire ST5 5BG, UK
[2] Institut de Recherches sur la Catalyse, 2, avenue Albert Einstein, 69626 Villeurbanne Cedex, France; *e-mail:* jobic@catalyse.univ-lyon1.fr

1 Introduction 31

2 Neutrons and Neutron Diffraction 32

3 Investigation of the Framework and Cations in Zeolites 38

3.1 Gallosilicates 38
3.2 Aluminosilicates 40
3.3 Aluminophosphates 42
3.4 Other Microporous Materials 44

4 Location of Adsorbed Hydrocarbon Molecules 47

4.1 Benzene in Faujasite Structures 47
4.2 Adsorption of Pyridine in Na-Y and Gallozeolite-L 49
4.3 Xylenes and Other Aromatics in Zeolites X and Y 50
4.4 Benzene in Potassium Zeolite L 54
4.5 Benzene in MFI Structure 55

5 Location of Small Physisorbed Molecules 56

6 Single-Crystal Studies on Natural Hydrated Minerals 57

7 Proton Positions and Hydronium Species 63

8 Concluding Remarks 66

References 67

1
Introduction

Neutron diffraction from single crystals and powders is a powerful technique for investigating the crystal structures of zeolites and microporous materials, as well as other inorganic compounds. Whereas X-ray diffraction is the method by which the framework structures of new zeolites are generally determined, either from a laboratory or synchrotron radiation source, neutrons can be used to probe in greater detail the arrangements of water molecules, hydrogen bonds and cations in the cavities, and the location of adsorbed molecules. The ordering of the tetrahedral atoms of the framework can also be investigated. Neutron diffraction is therefore complementary to other investigative techniques such as magic-angle-spinning NMR. Microporous materials, with the exception of some natural minerals, do not generally form good quality crystals of a size suitable for single-crystal neutron diffraction. Powder diffraction is therefore the technique that is most usually applied. However, for materials where single crystals are available, this is the preferred approach, because the absence of peak overlap greatly enhances the reliability of the refinements of complex structures. The investigation of zeolites by powder neutron diffraction was reviewed in 1986 by Newsam [1]. This article will review powder work from around this period onwards, as well as previous single-crystal studies. Other articles that are of interest which cover many of the practical aspects of powder diffraction include the reviews by Baerlocher [2], Baerlocher and McCusker [3], McCusker [4, 5], and Cheetham and Wilkinson [6]. The latter discuss many aspects of modern neutron and X-ray powder diffraction techniques. An overall introduction to neutron diffraction may be found in the books by Bacon [7] or Squires [8].

2 Neutrons and Neutron Diffraction

A neutron is an electrically neutral particle that has a spin of $\frac{1}{2}$ and hence a magnetic moment. Its rest mass (m_n) is 1.675×10^{-27} kg, so, by de Broglie's relationship ($\lambda = h/m_n v$) a neutron travelling with a speed (v) of 3956 ms^{-1} has a wavelength (λ) of 1 Å. This is of the same order of magnitude as the interplanar spacings in a crystal lattice, and hence is a suitable wavelength for carrying out diffraction experiments. The neutron is scattered by interaction with the nucleus of an atom via the strong nuclear force, or the magnetic moment can interact with unpaired spins via the electrostatic interaction. The nuclear scattering gives information about the arrangement of the atomic nuclei in the structure, whereas the magnetic scattering reveals the ordering of unpaired electron spin density. It is the nuclear scattering that is of most relevance in the investigation of zeolite structures.

A plane wave of neutrons incident on an isolated nucleus considered fixed at the origin is described by a wave function

$$\psi_{in} = \exp(i\kappa z)$$

where κ is the magnitude of the wave vector, i.e. $2\pi/\lambda$, and z is taken along the direction of travel of the neutrons towards the nucleus. The incident neutron flux is $v|\psi_{in}|^2$. After interacting with the nucleus, the scattered wave is spherically symmetrical and has the form

$$\psi_{sc} = -(b/r)\exp(i\kappa z)$$

where r is the distance from the nucleus, and b is the scattering length for the atom. If κ is the same before and after scattering, i.e. the wavelength is unaltered, no energy has been exchanged between the neutron and the nucleus, and the scattering is said to be elastic. The outgoing current of scattered neutrons integrated over all directions is given by $4\pi r^2 v|\psi_{sc}|^2$. The scattering cross section, σ, of the nucleus is defined as

$$\sigma = \frac{\text{outgoing current of scattered neutrons}}{\text{incident neutron flux}}$$

$$= \frac{4\pi r^2 v|(b/r)\exp(i\kappa z)|^2}{v|\exp(i\kappa z)|^2}$$

$$= 4\pi b^2 .$$

Hence the bigger the value of b for a nucleus, the greater the probability that an incident neutron will be scattered.

The nuclear scattering length, b, depends on the structure of the atomic nucleus and therefore varies, seemingly erratically, from one element to the next, and indeed from one isotope to another of the same element (Fig. 1). Within a factor of say 3 or 4, most atoms scatter neutrons equally well, which contrasts with the case of X-ray diffraction, where the scattering power increases steadily with the number of electrons in the atom. For example, uranium (element 92) scatters X-rays much more strongly than nitrogen (element 7), whereas with neutrons, nitrogen scatters 11% more strongly than uranium. Furthermore, because the range of the strong nuclear force ($\approx 10^{-15}$ m) is very small compared to the wavelength of the neutron, the neutrons are scattered spherically, i.e. there is no dependence of the scattering power on angle. This again contrasts with X-rays, where the electron cloud that scatters the photons is of a similar size to the wavelength; interference between the waves scattered from different parts of the electron density leads to a fall off in scattering power with angle, or form-factor (Fig. 2). With neutron diffraction, therefore, the intensities of the diffraction peaks decrease less rapidly with angle as compared to X-ray diffraction.

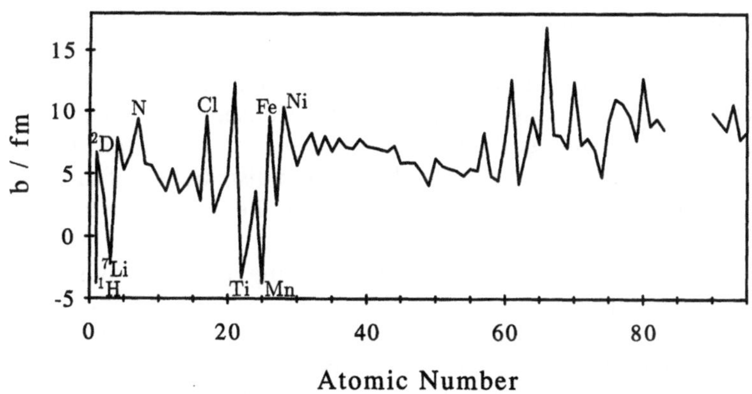

Fig. 1. Average neutron scattering lengths [10] for elements, and isotopes ^{1}H, ^{2}D and ^{7}Li

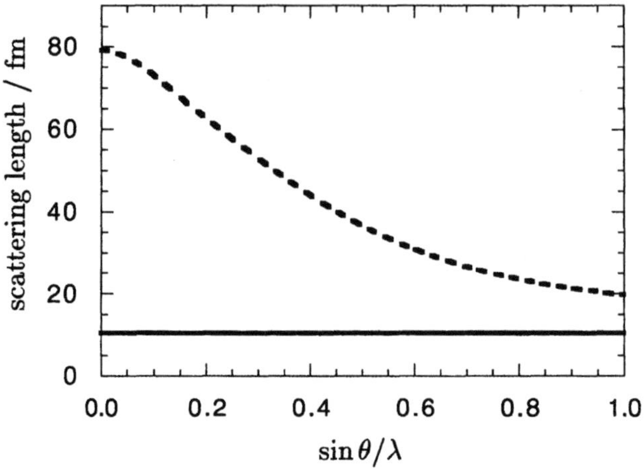

Fig. 2. Neutron (*solid line*) and X-ray (*dotted line*) scattering factors for nickel as a function of sin θ/λ. Neutrons are scattered isotropically, whereas with X-rays there is a form-factor fall off of scattering power with angle

For nuclei that have nuclear spin, the magnetic moment of the nucleus can interact with the magnetic moment of the neutron to modify the nuclear scattering length, depending on the spin state. Thus for normal hydrogen, ^1H, the proton has spin $\frac{1}{2}$ and hence two nuclear spin states of $+\frac{1}{2}$ and $-\frac{1}{2}$ occur. More generally, for a nuclear spin J there are $2J+1$ spin states. With hydrogen, the two spin states have scattering lengths of 1.04 fm and – 4.7 fm. These occur with equal probability, yielding an average scattering length of – 3.741 fm. Nuclei with positive and negative scattering lengths scatter neutron waves with opposite phases. For an assembly of nuclei, the nuclear scattering length varies randomly from one nucleus to the next, depending on the nuclear spin state. In a similar way, when a natural element has more than one isotope, the scattering length varies between nuclei depending on which isotope is actually occupying a particular position. These fluctuations in scattering length over an assembly of nuclei give rise to a coherent and an incoherent scattering cross section for the element, given by $\sigma_{coh} = 4\pi(\bar{b})^2$ and $\sigma_{inc} = 4\pi\{\bar{b^2} - (\bar{b})^2\}$, respectively. The coherent scattering depends on the mean value of the scattering lengths, $\bar{b}$, over the assembly, which is the same for all equivalent atomic sites. Consequently, it can give rise to interference effects, such as diffraction, and hence to structural information. The incoherent scattering depends on the random distribution of the deviations of the scattering lengths from their mean value. Being randomly distributed over the assembly, no interference effects can occur and hence no structural information is available from incoherent scattering. However, by studying the incoherently and inelastically scattered neutrons, information about the dynamics of the incoherent scatterers in an assembly can be obtained. In this respect, normal hydrogen, with its two spin states, has a very high incoherent scattering cross section, of 79.9×10^{-28} m^2, whereas its coherent cross section is much smaller, 1.8×10^{-28} m^2. Inelastic neutron spectroscopy is an important tool for studying the dynamics of systems containing hydrogen, including microporous materials [9]. Rotational and translational motions can be characterised by quasielastic neutron scattering (QENS). These motions occur at very small energy transfers and yield a continuous spectrum centred about the elastic peak [9].

The general uses of nuclear neutron diffraction include the location of light atoms, such as hydrogen, lithium, etc., which can scatter as strongly or more strongly than heavier elements, differentiation between neighbouring elements in the periodic table, for example the neutron scattering lengths of Al, Si and P are 3.449, 4.149, and 5.13 fm, respectively [10], and the characterisation of disordered systems since the absence of a form factor leads to much stronger scattering at high values of sin θ/λ, i.e. at small diffraction d spacings. The neutron is highly penetrating and is strongly absorbed only by those atoms with an appropriate nuclear transition, such as ^{10}B, ^{6}Li, ^{113}Cd, etc. Hence sample environments are easily constructed, and neutrons are useful for studying the behaviour of materials with temperature, pressure, or for following the course of a solid-state chemical reaction in a reaction vessel constructed from a low-absorbing material such as silica glass. When investigating systems containing hydrogen atoms, it is generally desirable, if possible, to use a deuterated form of the sample, to avoid the high incoherent background due to ^{1}H, which is detrimental to the statistical quality of the data.

Neutrons for diffraction experiments are produced by the fission of uranium in a nuclear reactor, or by the bombardment of a heavy metal target (e.g. uranium or tantalum) by an energetic beam of particles, such as protons. The former gives a steady-state source whereas the latter can give either a steady source or a pulsed source depending on the time structure of the proton beam hitting the target. The neutrons are emitted from the nuclei with very high energies, and must be at least partly moderated, by exchanging thermal energy with their surroundings or with hydrogenous materials such as liquid methane placed in their path, so that they have a distribution of energies appropriate for the diffraction experiment.

From the source, the neutrons are led to the experimental area in guide tubes. With a steady-state source, neutrons with a narrow distribution of wavelengths ($\Delta\lambda$) are selected using Bragg diffraction from a crystal monochromator. The monochromatic beam is incident on the sample, and the scattered neutrons are detected by scanning a bank of detectors around the sample position. Compared to even a conventional X-ray source the flux from a reactor is quite low. Hence the highest flux at the sample, the maximum detector efficiency, and a relatively large volume of sample are required to ensure reasonable counting times. A typical scan to collect data with a high-resolution powder diffractometer may last several hours. As an alternative to scanning a bank of single detectors, position-sensitive detectors can be used, but these tend to have poorer angular resolution and are used principally for dynamic measurements, rather than high-resolution experiments for the characterisation of complex crystal structures.

With a pulsed source of neutrons the time-of-flight method (TOF) can be used. Here the detectors are placed at fixed scattering angles (2θ) around the sample, and the time taken for each neutron to travel from the source to the detector is recorded. Since a neutron's speed is related to its wavelength by de

Broglie's relationship, the wavelength of each neutron can be calculated, and the spacing d_{hkl} between the reflecting lattice planes in the sample obtained from Bragg's law, $d_{hkl} = \lambda/2 \sin \theta$.

Both methods have their advantages. With the TOF method the whole powder diffraction pattern is collected simultaneously, and because the moderation of the neutrons is incomplete, to maintain a temporally compact pulse, there are many short-wavelength neutrons available so that data can be easily collected to high values in $\sin \theta/\lambda$. The resolution of the instrument can also be made superior in this regime by having a long flight path and detector banks at high angle, which has advantages for the accurate refinement of crystal structures. By contrast, there are fewer long-wavelength neutrons available, making measurement of the large d-spacings from a specimen more difficult. This is more easily accomplished using longer wavelength neutrons at low 2θ values such as are available using a reactor source. With monochromatic neutrons the treatment of the data is much simpler because there is no need to make energy-dependent corrections such as absorption, or to correct for the variation of the incident flux with neutron energy.

Whether using monochromatic radiation, or the TOF technique, the diffraction pattern consists, in the single crystal case, of a set of intensities for reflections from the various lattice planes of the crystal. For a powder, the diffraction pattern is recorded as intensity vs. scattering angle 2θ for monochromatic radiation, or vs. d spacing or TOF for pulsed radiation. The intensity of the radiation scattered from a set of crystal lattice planes is dependent on the structure factor F_{hkl} which is determined by the arrangement of the atoms within the unit cell of the crystal, i.e.

$$F_{hkl} = \sum_{i=1}^{N} b_i \exp\{2\pi i(hx_i + ky_i + lz_i)\} \exp\left\{-\frac{1}{4} B_i \left(\frac{1}{d_{hkl}}\right)^2\right\}$$

where the summation is over all the N atoms in the unit cell at positions (x_i, y_i, z_i). The term $\exp\left\{-\frac{1}{4} B_i \left(\frac{1}{d_{hkl}}\right)^2\right\}$ is the isotropic Debye–Waller factor which takes account of small displacements, either static or due to thermal motion, of the atoms about their mean positions in the structure. The larger the value of B_i, which is usually referred to as the isotropic temperature factor, the greater the root mean square displacement of the atom i about its mean position.

To refine the structure from single-crystal data, the various parameters that characterise a crystal structure, e.g. atom positions, occupancies, Debye–Waller factors, etc., are allowed to vary so as to obtain the best least-squares agreement between the observed reflection intensities and those calculated from the structural model. In the powder case, there is usually overlap between adjacent peaks, so that the individual intensities are no longer measurable. Nevertheless, as illustrated in Fig. 3, the overlapping peaks produce a complex diffraction profile that clearly contains a lot of structural information. The refinement of a crystal structure from powder data makes use of the Rietveld method [11]; the complex diffraction profile is calculated from an initial structural model as a sum of overlapping reflections, and compared to the observed pattern. To do this a peak-

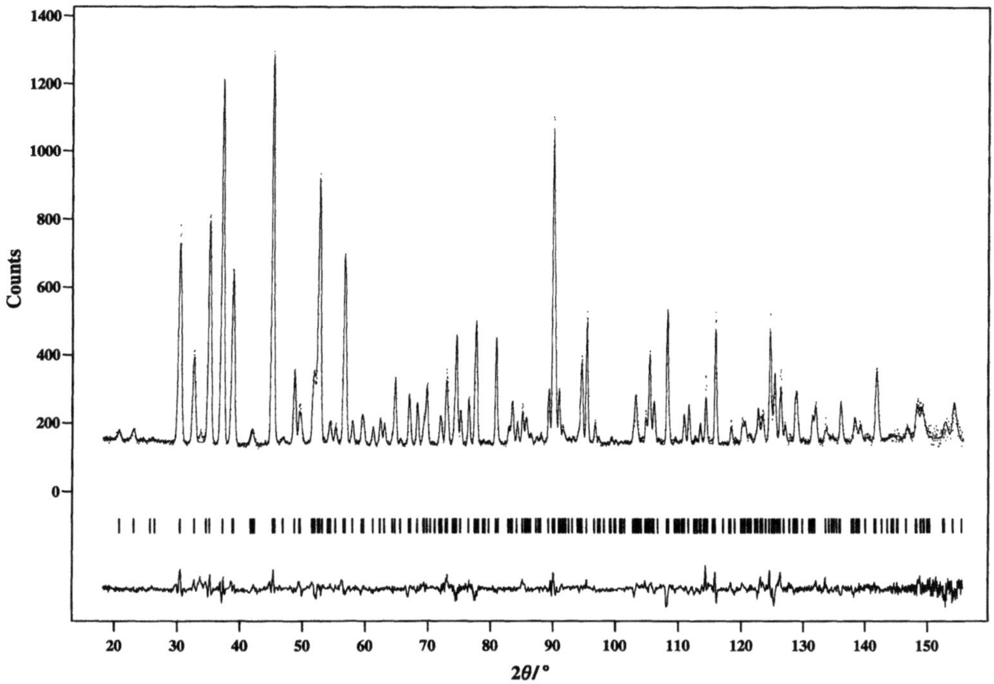

Fig. 3. A powder neutron diffraction pattern measured with a monochromatic beam of neutrons. The *points* represent the counts measured on the diffractometer, the *solid line* is the diffraction profile calculated from the best structural model fitted by least-squares using the Rietveld method. The *bottom solid line* is the difference between the observed and calculated profiles. Ideally, with a perfect model, differences between the observed and calculated profiles should be due only to counting statistics. In reality, there are always deficiencies in a model (both in the structure and the description of the peak and profile parameters) leading to additional features in the difference curve. *Vertical lines* indicate the 2θ positions of each of the reflections contributing to the profile. As can be seen, especially at high angle, peak overlap is extensive

shape function must be assumed. For monochromatic data, this is usually a Gaussian, or more likely, a pseudo-Voigt function, which is the normalised sum of a Gaussian and a Lorentzian component. For TOF data more complex functions are required. The positions of the peaks are calculated from the unit-cell parameters of the model structure, the peak widths have a simple dependence on 2θ or d, and the peak intensities are determined by the arrangement of atoms in the crystal. By varying the unit-cell parameters, the parameters that describe the variation of peak width with angle or d, and the atomic parameters, the best least-squares fit between observed and calculated diffraction profiles can be obtained, and hence the best crystal structure is refined. Figure 3 shows the result of a typical Rietveld refinement from monochromatic powder neutron data.

3 Investigation of the Framework and Cations in Zeolites

Early work using powder neutron diffraction was mainly devoted to studying variations of the framework caused by cation exchange, changes in composition, temperature, dehydration etc., and the nature (if any) of ordering of Si and Al. Examples can be found in reviews [1, 12].

3.1 Gallosilicates

A hydrated lithium gallosilicate with the ABW framework [13] $LiGaSiO_4 \cdot H_2O$ was investigated at 19 and 298 K [14, 15] and was shown to have strict alternation of Si and Ga in the framework, as might be expected from Loewenstein's rule [16]. One lithium-ion site and one water molecule were located. The lithium is tetrahedrally coordinated by three framework oxygens and the oxygen of the water molecule. The positions of the hydrogen atoms indicate only weak hydrogen bonding to a framework oxygen and an adjacent water molecule. It is clear that the position of the water molecule is dominated by the interaction with the lithium ion, and that the hydrogen bonding is of secondary importance (Fig. 4). The Si–O bond lengths are slightly longer than normally expected [mean distance of 1.647(5) Å] which is believed to arise from the influence of the lithium ions, the acute nature of the average Si–O–Ga bond angle, and the direct electronic effect of having Ga in the framework rather than Al.

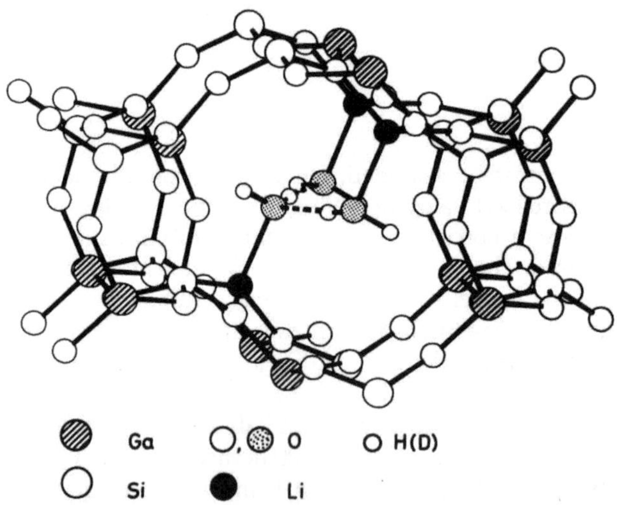

Fig. 4. View along the channel of the ABW gallosilicate showing the framework ordering, the lithium cation position, the position of the adsorbed water molecule, and the weak hydrogen bonding [14]

For the analogous aluminogallosilicate, $LiAl_{0.5}Ga_{0.5}SiO_4 \cdot H_2O$, a similar structure was seen [17]. There is no evidence for the formation of a supercell, or for a reduction in the symmetry, which would be required if there were ordering of the Ga and Al atoms. Hence Loewenstein's rule applies, with the segregation of Si and Al/Ga onto separate sites, but on the Al/Ga site the arrangement is disordered on the long-range crystallographic scale. A single Si–4(Al/Ga) peak in the ^{29}Si NMR spectrum is also consistent with a disordered Ga/Al distribution.

Hydrated and dehydrated synthetic sodium-cesium gallosilicates, with higher Si/Ga ratios in the range 2.45–2.49 and the ANA framework [13], were found to have compositions comparable to those of the natural aluminosilicate pollucite, with Ga replacing Al [18]. No evidence was found for long-range ordering of Si and Ga which were assigned a single crystallographic site. For the hydrated specimens, the positions of the water molecules were consistent with an early neutron single-crystal study on analcime [19], with water molecules occupying the same site at (1/8,1/8,1/8) as Cs, to give overall full crystallographic occupancy. There is little change in the framework on dehydration, although a small shift in the Na position is seen, to accommodate the loss of the partly coordinating water molecules which are at a distance of 2.43 Å in the hydrated samples.

The structures of three gallosilicates with the FAU framework [13], $Na_xGa_xSi_{192-x}O_{384}$, (with $x = 85, 72$ and 57, Si/Ga = 1.25, 1.67 and 2.36, corresponding to aluminosilicate zeolites Na-X, an intermediate composition, and Na-Y), were refined [20]. The space group was consistently found to be $Fd\bar{3}m$ with long-range disorder of Si and Ga over one crystallographic site, as opposed to the lower symmetry space group $Fd\bar{3}$, which would result if there were two sites for the framework tetrahedral atoms because of Si and Ga segregation. The occupancies of various sodium-ion sites were refined, but only four were found to be relevant for this system, i.e. sites I, I′, II, and III [21]. Site II in the supercage is always fully occupied. The occupancies of site I, in the centre of the hexagonal prism, and site I′, in the six-ring window just inside the sodalite cage next to the hexagonal prism, appear to adhere to the notion that occupancy of site I excludes occupancy of the two adjacent site I′ positions because of their proximity. The maximum occupancy for these sites is thus given by 2 × site I + site I′ = 32. Hence if the total occupancy of sites I and I′ is less than 32, then, for a given hexagonal prism, either the SI site is occupied or the *two* adjacent SI′ sites. As the number of sodium ions increases with x, so the occupancy of site I′ increases at the expense of site I, so that in zeolite X, site I′ is fully occupied with 32 ions per unit cell. This trend is consistent with the occupancy of sites I, I′ and II seen for Na-Y [22]. Site III appears to accept the overspill that cannot be accommodated in sites I, I′ and II. It seems likely that the way the occupancies of the cation sites evolve with composition is responsible for the discontinuities seen in certain framework properties in FAU zeolites – such as lattice parameter – rather than effects associated with long-range coherence in the local Si/Al ordering schemes.

3.2 Aluminosilicates

The ionic conductivities of dehydrated alkali-metal zeolites X and Y have been studied [23]. Substitution of a small amount of Li^+ into Na-X causes a large decrease in conductivity below 150 °C. However, the conductivity of Li-X is higher than that of Na-X above 350 °C. A powder neutron diffraction study of Li-X and Li-Y, with a very high level of Li exchange, revealed two sites for Li in Li-Y, at sites I′ and II, and for Li-X, a third position at site III′ was also found. The negative scattering length of Li for neutrons means that the first two sites were seen as negative peaks in the Fourier maps. The third site for Li-X was located via a systematic investigation of possible positions of the cation in the refinement. The SII site is fully occupied in Li-X, but only 67% occupied in Li-Y, which contrasts with the situation in Na-X and Y, where the SII site always appears to be full. The site III′ is located in a mirror plane near the twelve-ring window. The lithium ion is coordinated on one side by two O(4) atoms at 2.13(5) Å, two O(1) at 2.30(5) Å, and one O(3) at 2.22(8) Å.

Synthesis of a dealuminated zeolite Y of a highly crystalline "zero defect" nature was evident from the sharp diffraction peaks seen in a time-of-flight neutron diffraction study [24]. Dealuminated (or ultrastable) zeolite Y is of particular importance in the petroleum cracking industry because of its enhanced thermal stability. The sample was synthesised by treating Na-Y with silicon tetrachloride vapour, and was then washed and heated to remove aluminium that may have accumulated in the pores. Refinement of the occupancy of the tetrahedral framework atoms showed no significant deviation from being entirely silicon. No aluminium in the framework or sodium ions in the channels are therefore present. Attempts to locate adsorbed benzene molecules at a coverage of around one molecule per supercage at 4 K were not successful. Although the increase in lattice parameter with respect to the bare zeolite confirms benzene is in the sample, the absence of any cation sites implies a very flat potential energy surface so that the benzene is highly disordered in the channels, distributed over a range of low symmetry sites greatly reducing its influence on the Bragg intensities. Deuterium NMR studies of this system indicate that the behaviour is different from that observed in Na-Y (*vide infra*).

Zeolite ZK-4 is a more silicious analogue of zeolite A. Zeolite A was the subject of much study in the early 1980's by powder neutron diffraction and solid-state NMR to assess the nature of the framework ordering [1, 12]. With a Si/Al ratio very close to unity, Loewenstein's rule is expected to hold, with strict alternation of Si and Al in the framework. Hence the structure is described in the cubic space group $Fm\bar{3}c$ with a lattice constant of approximately 24 Å. For sodium ZK-4, with a Si/Al ratio of 1.65, the absence of superstructure peaks indicate that the structure can be described in a subcell, in space group $Pm3m$, with a lattice parameter of 12.19 Å [25]. This implies that there is no long-range ordering of Si and Al, though local ordering is no doubt present, as suggested by ^{29}Si MAS NMR. The sodium ions were located in the expected position in the six-ring window. Additional peaks in the difference Fourier maps suggest some residual material from decomposition of the organic template, or slight dealumination of the framework.

Measurements by NMR, IR, and neutron diffraction of deep-bed calcined zeolite D-ZK-5 show that calcination at 500 and 650 °C causes dealumination of the framework, and the loss of bridging OH groups [26]. Methanol adsorption is 75 and 45% of the theoretical maximum for calcination at 500 and 650 °C, respectively. Cesium ions in the sample are found at the centre of a single eight-ring window. A difference Fourier map reveals density that appears to be the Al and O atoms of aluminium removed from the framework. The occupancy of this site is higher in the sample calcined at the lower temperature. This is possibly because at higher temperatures the aluminium removed from the framework forms more randomly oriented clusters, thus reducing their prominence in the Fourier maps.

Detailed studies of the framework changes and the location of non-framework aluminium following calcination have been reported for zeolite D-Rho. The studies involve deep-bed calcination at 500 and 700 °C [27], shallow-bed calcination at 500 and 600 °C [28], and shallow-bed calcination in steam at 500 °C [29, 30]. Deep-bed calcination results in extensive dealumination, whereas shallow-bed results in less damage. With steam, around half the framework-Al is removed, and a weakly acidic bridging hydroxyl group can be identified.

Zeolite Rho is renowned for the flexibility of its framework, which responds to changes in the level of hydration or cation exchange. This offers perhaps the prospect of tailoring shape selectivity in catalytic processes by correctly siting cations to cause the appropriate framework distortion, e.g. changes to the aperture of the openings connecting the cages. In situ X-ray diffraction was used to examine changes to various samples of zeolite Rho containing divalent cations (Ca^{2+}, Sr^{2+}, Ba^{2+}, Cd^{2+}) as a function of temperature [31]. Mixed Ca^{2+}/Cs^{+} and Ca^{2+}/NH_4^{+} samples were also investigated. For the latter, the effects of dehydration (at ≈ 100 °C) and the evolution of ammonia (at ≈ 400 °C) were clearly evident. Neutron diffraction studies of Ca^{2+}/ND_4^{+} exchanged samples show that at room temperature there are two coexistent phases, with either a high ND_4^{+} and low Ca^{2+} content, or vice versa. Structural parameters were refined for both phases by a multiphase Rietveld refinement. The sample with the higher Ca content has the smaller lattice parameter, because the polarising Ca^{2+} ion causes more distortion to the zeolite framework. After deammoniation by heating to 400 °C, the two-phase mixture becomes single phase, with a redistribution of calcium ions to form Ca/D-Rho. Comparison of the frameworks of H-Rho with that of Ca/D-Rho shows a very significant distortion to the eight-rings which form highly anisotropic ellipses. The distortion leads to a 21% decrease in unit-cell volume, illustrating the enormous flexibility of the framework [32] (Fig. 5).

The framework distortion and the location of Rb and water molecules were compared in the dehydrated aluminosilicate $Rb_{10}Al_{10}Si_{38}O_{96}$ and the novel berylloarsenate $Rb_{24}Be_{24}As_{24}O_{96} \cdot 3.2D_2O$ [33]. Both have the Rho framework. Whereas the structure of the aluminosilicate could be easily refined from powder neutron diffraction data, yielding two Rb sites, a combined refinement with X-ray diffraction data was required for the berylloarsenate in order to provide greater contrast between Rb and water, as the scattering powers of oxygen and hydrogen for X-rays are very much lower than Rb. Four Rb and three water sites were obtained, and an elliptical distortion to the framework eight-

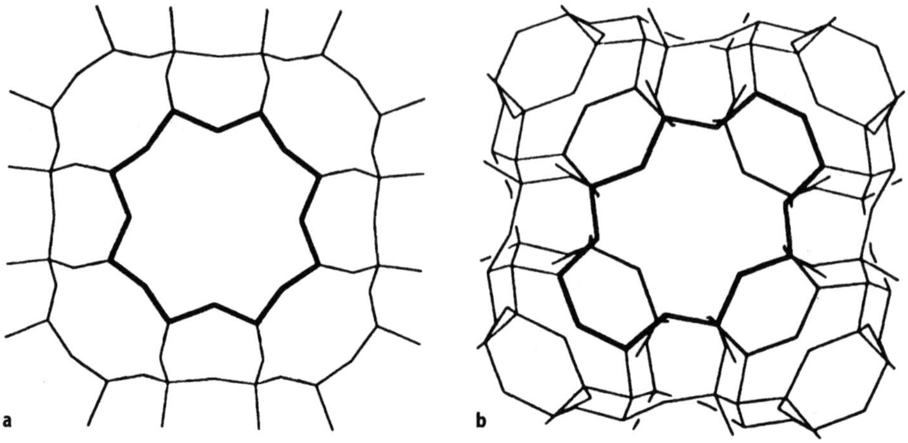

Fig. 5 a, b. Projection of the zeolite Rho framework for **a** D-Rho ($a = 15.0976(4)$ Å) and **b** Ca/D-Rho ($a = 13.9645(7)$ Å), illustrating the enormous flexibility of the Rho framework [32]

rings was seen that appeared to follow the same trend with lattice parameter established for other materials with the Rho framework. Further work has used neutron and X-ray powder diffraction to characterise various Rb- and Tl-substituted compounds with beryllophosphate, berylloarsenate and aluminosilicate frameworks [34].

A highly silicious analogue of ferrierite has been investigated by synchrotron and neutron powder diffraction, ^{29}Si MAS NMR, and energy minimisation calculations [35]. Ferrierite is reported to have excellent shape selectivity in the catalytic isomerisation of n-butene to isobutene. The crystal structure of natural ferrierite is reported in the high symmetry space group *Immm*. However this imposes a 180° T–O–T bond angle, which may indicate that the symmetry is in reality lower, thus allowing a relaxation from this value. Both the X-ray and neutron refinements produce better fits in space group *Pmnn*, a subgroup of *Immm*, with weak reflections in the X-ray data that are forbidden in *Immm* confirming the lower symmetry. The final refinements were made simultaneously to both X-ray and neutron data sets, to produce the most precise determination of the atomic parameters. The 180° bond angle relaxes to 170° in the lower symmetry description. Atomic positions calculated from empirical force fields for silicates produced excellent agreement with the experimental structure. Furthermore, the ^{29}Si NMR spectrum cannot be fitted assuming four distinct Si atoms (in the ratios 4:8:8:16) as required for the model in space group *Immm*, but can be fitted qualitatively using the five resonances of the lower symmetry *Pmnn* model, provided they are allowed to have different line widths.

3.3
Aluminophosphates

Aluminophosphates (or AlPOs) are microporous materials built from vertex-sharing AlO_4 and PO_4 units in a way analogous to aluminosilicate zeolites. Whilst some of these compounds and their substituted derivatives have the same structures as zeolites, others have structures for which there is no aluminosilicate equivalent. From a chemical viewpoint, strict alternation of Al and P in the framework is to be expected. Because Al and P have different scattering lengths, neutron diffraction should be a sensitive probe of this ordering in the framework, whose occurrence can also be confirmed from ^{31}P and ^{27}Al MAS NMR.

For AlPOs, powders with a high degree of crystallinity have often been hard to synthesize. Hence peak broadening in the powder diffraction patterns is usual. With large unit cells, and complex structures, there are many overlapping reflections, and this problem is exacerbated by the peak broadening. The alternation of Al and P in the framework leads to strong pseudosymmetry, with a close resemblance to a disordered structure without Al/P ordering. Under these conditions it is very difficult to refine an ordered structural model, even from neutron diffraction, and several studies have been limited to a description of the overall average structure in a higher symmetry space group than that required for Al/P ordering.

Nevertheless important structural information has been deduced from neutron powder diffraction studies on AlPOs. Thus the structure of $AlPO_4$-11 was solved from a combination of methods [36]. From electron diffraction the unit cell and lattice type were deduced, and were found to be consistent with X-ray powder diffraction patterns. The framework topology was obtained from a modelling technique, and revealed a new structure, related to $AlPO_4$-5 by removal of one set of four-rings. The refinement of the structure was carried out from time-of-flight powder neutron diffraction data in space group *Icmm*, with Al and P disordered in the framework. The refinement confirmed the proposed framework structure. Ordering of the Al and P in the framework requires a reduction in symmetry to *Icm*2. Refinements were attempted in this space group [37] and appear to confirm the ordered structure. However, because of the complexity of the ordered model, problems with the convergence of the refinement were encountered, and the ranges in bond lengths and angles are larger than expected.

In a similar manner, the highly complex structure of $AlPO_4$-54, or VPI-5, was determined from powder diffraction data [38]. The unit cell and hexagonal symmetry suggested a similarity to the (then) theoretical network 81, which has an eighteen-ring channel. The structure was refined from neutron powder data, without framework ordering in the space group $P6_3/mcm$; again the proposed structure, the first of its kind, was confirmed. High resolution X-ray powder diffraction studies using synchrotron radiation and a highly crystalline sample showed that the symmetry must be reduced to space group $P6_3$, which is lower than that required for simple ordering of P and Al in the framework (space group $P6_3cm$), leading to six crystallographically distinct T-atoms in the framework. The water molecules were also located running in a triple helix of hydrogen-bonded chains down the eighteen-ring channel [39]. The relationship between $AlPO_4$-8 and VPI-5, from which $AlPO_4$-8 can be prepared by calcination, led to the solution of the latter's structure, in conjunction with evidence from electron microscopy and model building [40]. Because of the problems posed by a complex structure with pseudosymmetry, the refinement from neutron diffraction data was conducted in the higher symmetry, disordered space group.

One of the simplest microporous aluminophosphates is $AlPO_4$-5. Single-crystal X-ray diffraction determinations of the structure before calcination showed hexagonal symmetry, with strict alternation of Al and P in the framework (space group *P6cc*) [41, 42]. A refinement of the structure of calcined material from powder neutron diffraction data did not provide a convincing answer, and a satisfactory fit was only achieved in the higher symmetry space group *P6/mcc* where again there is disorder of the Al and P [43]. This result was unexpected as it conflicts with the evidence of the single-crystal X-ray studies.

The problem arises because during calcination the framework undergoes a small distortion to an orthorhombic structure. The distortion is revealed by tiny splittings to diffraction peaks that are seen from a well-crystallised sample using high resolution X-ray powder diffraction of synchrotron radiation [44]. The presence of other very weak peaks shows a further reduction in symmetry. A number of models in different subgroups of the hexagonal space groups were investigated, and that in *Pcc*2 was found to give the best fit to both the X-ray data, and a new powder neutron diffraction pattern taken of the same well-crystallised sample. The neutron data, with the greater contrast between Al and P, ought to be a more stringent test of the different models. The final model adopted was that refined from the neutron data (Fig. 6). A better refinement was obtained because of the higher quality of the neutron data out to small *d*-spacings, and because the enhanced contrast between Al and P leads to a reduction in the pseudosymmetry as compared to the X-ray pattern. However, the angular resolution of the neutron data was much lower than that of the X-rays, and without the latter, the reduction in symmetry from hexagonal to orthorhombic would have been very hard to see. As with ferrierite above, there is much to be gained when studying complex microporous systems by using a number of complementary techniques.

3.4
Other Microporous Materials

The ingenuity of the chemist knows no bounds, and more and more microporous materials are being synthesised using elements from throughout the periodic table. Where these cannot be grown as single crystals, powder methods must be used for structural solution. Whereas X-ray data is often the method of choice for an initial determination of the structure, neutron diffraction often plays its role to investigate framework-atom ordering, to locate light atoms, disordered cations and adsorbed water molecules, to investigate hydrogen bonding, and to provide good quality structural refinements. It is generally true that the accuracy of structural refinements from powder neutron data is often higher than that from X-ray data, because the data extends to lower *d*-spacings, and there are fewer problems with absorption, and preferred orientation with a large sample.

Some examples where neutron diffraction has contributed to studies of novel microporous materials include studies of $LiZnPO_4 \cdot H_2O$, where the positions of

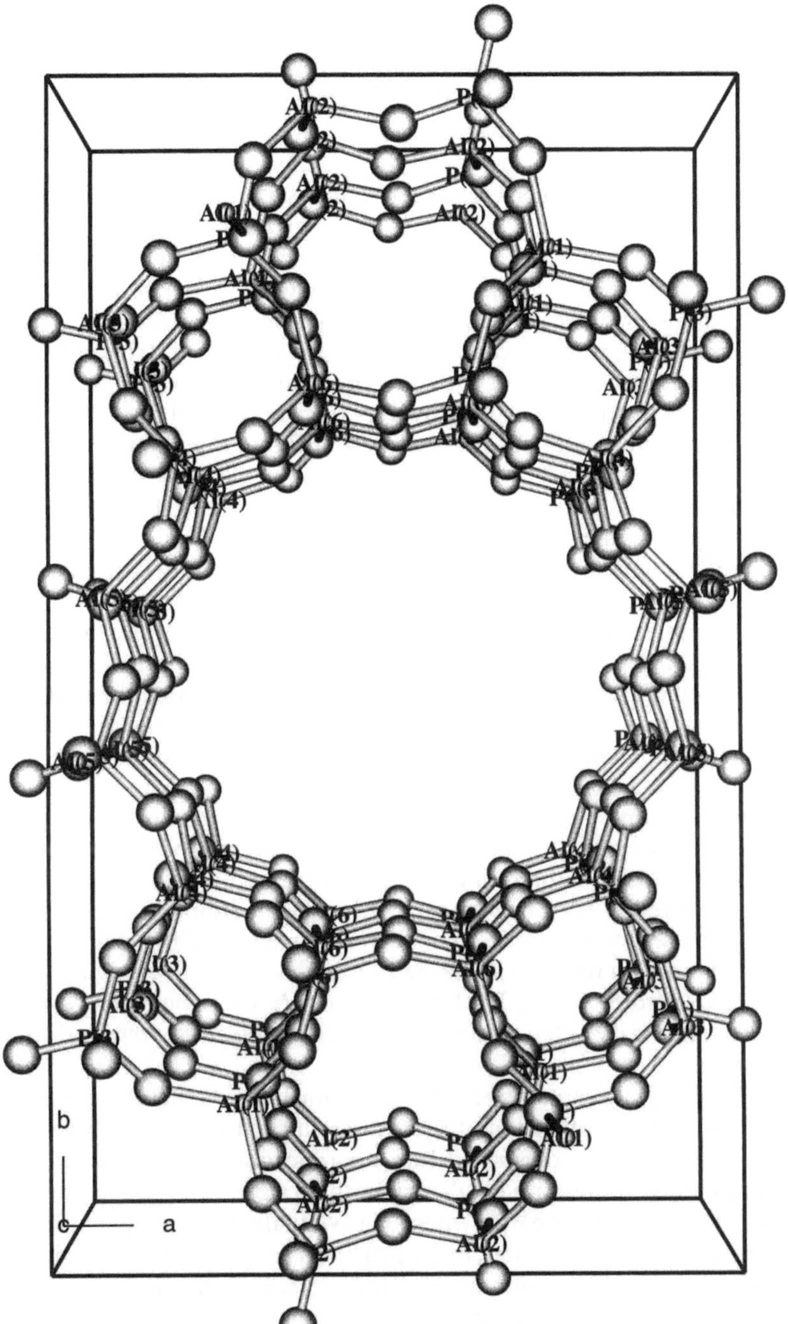

Fig. 6. View down the *c*-axis of $AlPO_4$-5 showing the Al and P ordering in the orthorhombic space group *Pcc*2. In this space group there are two crystallographically distinct twelve-ring channels

lithium, water and the hydrogen atoms are located in the cavities [45]. The framework is of the type ABW, and the Li cation and the water molecule take up positions as in the analogous gallosilicate $LiGaSiO_4 \cdot H_2O$ discussed above [14, 15]. A dehydrated magnesium-sodium beryllophosphate, $Mg_{19}Na_{58}(BePO_4)_{96}$, was found to have the faujasite structure [46]. Strict alternation of Be and P in the framework reduces the symmetry from $Fd\bar{3}m$ found in zeolite Y to $Fd\bar{3}$. Magnesium and sodium ions were ordered onto separate sites. Mg is at the centre of the hexagonal prism (SI) and Na is found on its preferred position in zeolites X and Y, the SII site. The structure of a sodium zinc arsenate, $Na_6(ZnAsO_4)_6$, was solved by simulated annealing from X-ray powder diffraction data. However the final refinements were made from low-temperature neutron diffraction data [47]. The framework is of the condensed hexagonal "stuffed" tridymite structural type and undergoes a reversible transformation on rehydration to a framework analogous to sodalite, which involves the breaking of T–O–T bonds and their reforming into a different framework, whilst still maintaining Zn/As alternation. There is no equivalent transformation in the aluminosilicate analogue.

The structure of a complex, non-centrosymmetric gallophosphite, $Ga_2(HPO_3)_3 \cdot 4H_2O$, was solved from a combination of high-resolution synchrotron radiation diffraction and powder neutron diffraction [48]. Twenty-nine distinct atoms were located in the asymmetric unit of the crystal structure. The overall structure was solved using direct methods and Fourier synthesis from the X-ray data, to reveal the positions of 17 atoms. Hydrogen-atom positions were obtained from subsequent analysis of the neutron data, in which the fourth water molecule was found, whose presence was not expected following the initial thermogravimetric analysis of this new compound. The structure is made up of units comprising two GaO_6 octahedra linked by three phosphite groups through shared oxygen atoms. The units are connected into a three-dimensional framework by sharing oxygens with other units and by hydrogen bonding. This produces small elliptical channels lined with hydrogen atoms. The X-ray data cannot be used to reveal the arrangement of hydrogen atoms in this compound, and the structure is too complex to be solved by neutron diffraction data alone. Splitting the problem into two manageable stages, structure solution and heavy-atom structure from X-rays, light atom structure from neutrons, greatly increases the complexity of the structures that can be solved by powder diffraction techniques.

The gallophosphite contains octahedrally coordinated Ga atoms in the framework. Structural units other than tetrahedra will become more common as the diversity of elements incorporated into microporous compounds progresses. Other compounds containing octahedra in the framework are the members of the series $M_{4-x}H_xGe_7O_{16} \cdot nH_2O$, where M is virtually any univalent cation. The arrangement of water molecules, hydrogen bonds and cations in the silver-substituted (Ag_4) and sodium-ammonium (Na_3NH_4) forms were also obtained from a combination of high-resolution powder X-ray and neutron diffraction [49]. The strongly scattering Ag^+ ions were easily located in two sites by the X-rays, and the water molecules and hydrogen atoms were found sequentially by analysis of the X-ray and neutron data. Nitrogen is a strong scatterer of neutrons (with a scattering length of 9.36 fm) and the neutron data can be used to confirm the replacement of one of the Ag^+ sites by ND_4^+, and the other by the three Na^+ ions. In these two isomorphous compounds, the water molecules and cations form two-dimensional sheets that run perpendicular to the body diagonal of a rhombohedral unit cell that is only slightly distorted from cubic.

4 Location of Adsorbed Hydrocarbon Molecules

Molecules adsorbed into zeolites are very often disordered. The disorder can arise from there being more than one chemically distinct site for the location of the molecules, or because the high symmetry of some zeolites means that there are many symmetrically equivalent sites for the molecules. For steric reasons, not all sites can be occupied simultaneously. The forces between the adsorbed molecules and the framework or interchannel cations are generally quite weak, leading to considerable thermal motion of the molecules at ambient temperatures. Hence there are frequently a number of partially occupied sites for adsorbed molecules with relatively facile exchange between them. Neutrons are therefore well suited to investigating these systems. The absence of a form factor means that diffracted intensities are stronger at high angle than is the case with X-rays, leading to a more reliable distinction between the effects of partial occupancy and the effects of thermal motion. For hydrocarbon (preferably deuterated) molecules, the hydrogen atoms make a large contribution to the scattering, thus improving the chances of accurately locating the molecules.

4.1 Benzene in Faujasite Structures

One of the earliest investigations of the position of adsorbed hydrocarbon molecules in zeolites by powder diffraction was the location of deuterated benzene in dehydrated sodium zeolite Y (Si/Al = 2.43) [12, 22, 50, 51]. Three samples containing no benzene, approximately 1.1 and 2.6 molecules per supercage, were studied at ambient temperature and at 4 K. After analysis of the data from the bare zeolite, which revealed the presence of three sites for the sodium ions in agreement with previous X-ray studies [52], this arrangement was used as a starting model for the data from the sample containing the lower coverage of benzene at 4 K. Fourier difference maps revealed the presence of the first benzene position, which lies perpendicular to the body diagonal of the cubic unit cell at an average distance of 2.70(1) Å from the SII sodium ions in the supercage. This sodium-ion site is fully occupied and is coordinated to six oxygen atoms forming the hexagonal face of a sodalite cage. There is clearly an interaction between the polarising sodium ion and the π-electron density of the benzene stabilising the molecule in this position.

At higher coverage, a second benzene site was found in the twelve-ring window between supercages. As well as an interaction between the hydrogen atoms of the molecule and the oxygen atoms of the window, there are favourable distances to the hydrogen atoms of neighbouring benzene(1) sites and these also help stabilise this position in the window. A small occupancy of the second site was subsequently refined at lower coverage, and it seems that the proportion of benzene in the second site increases with overall coverage, supporting the notion that this site is stabilised by interaction with occupied benzene(1) sites. Small angle neutron scattering studies of benzene in Na-Y [53] indicate that at low coverages (1 molecule per supercage) the benzene molecules are distributed rather evenly throughout the channels, whereas at higher coverages (2.5 molecules per supercage), clusters of molecules with a size of around 20 Å form. The arrangement of molecules in such clusters can be easily envisaged from the two benzene positions found by neutron diffraction. Each supercage has four benzene(1) sites and there is sufficient room for all four to be occupied simultaneously. If four molecules in each of two adjacent supercages are linked by a molecule in the window site, then a compact cluster with the appropriate dimension is formed (Fig. 7). Small shifts in the positions of the sodium ions and the centres of the benzene molecules at higher coverage seen in the diffraction experiment are compatible with the formation of such clusters. The room-temperature studies showed that the molecules are in similar positions to those at 4 K, but the high temperature factors indicate they undergo vigorous molecular motion.

Theoretical papers [54, 55] appear to reproduce well the behaviour determined by neutron diffraction. NMR investigations of the distribution of the molecules throughout the zeolite cavities [56–60], and spectroscopic measurements [61, 62] are in good agreement.

Similar positions are seen for the location of benzene in H-SAPO-37 [63], the silicoaluminophosphate equivalent of zeolite Y with Al/Si/P ratios of 1.0:0.31:0.69. At a coverage of one molecule per supercage, the preferred adsorption site is in the twelve-ring window (22% occupancy) with about half

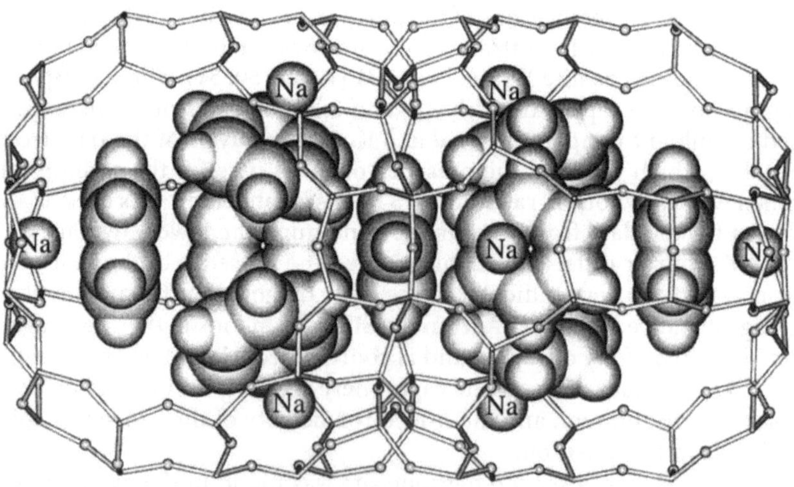

Fig. 7. Two adjacent supercages in sodium zeolite Y, showing a cluster of adsorbed benzene molecules, formed from 8 benzene(1) molecules bonded to the SII sodium ions in the two supercages, linked by a benzene(2) molecule in the twelve-ring window

as much (12% occupancy) adsorbed above the six-ring window. In this position the benzene has two orientations about its molecular sixfold axis, rotated by 30° with respect to each other, and is stabilised by a facial interaction with the hydrogen atom attached to the six-ring oxygen atom O(2), whose location is similar to that of the SII cation in Na-Y. Prior to adsorption of the benzene, the neutron diffraction data from the bare zeolite confirm the occurrence of Al/P ordering in the framework, with Si substituting for P, and show that hydrogen atoms are attached to O(1), O(2) and O(3), with occupancies of 20, 6 and 4%, respectively. However, after adsorption of benzene, hydrogen is found attached only to O(2). Hence hydrogen atoms are redistributing themselves to interact more favourably with the sorbate molecules. This deduction is compatible with the accompanying infrared measurements. The authors postulate, based on the observed similarity of the diffusion of benzene in H-Y, H-SAPO-37 and ultrastable-Y (determined from NMR T_1 measurements), that proton migration to the O(2) site might occur in all these systems. This would imply that the Brønsted acid activity of a zeolite catalyst may be determined to some extent by the sorbate, and the extent to which favourable interactions can be formed between the sorbate and the acid sites.

4.2 Adsorption of Pyridine in Na-Y and Gallozeolite-L

Spectroscopic studies indicate that pyridine in Na-Y is adsorbed at a Lewis acid site [64, 65]. A neutron diffraction study at low coverage [66] suggests that the only adsorption site for pyridine is centred in the twelve-ring window, very similar to the position of benzene(2). The pyridine molecule has sixfold orientational disorder with the nitrogen atom pointing towards the O(4) atoms of the window. The apparent nitrogen-to-oxygen distances of 3.84 Å [to O(4)] and 3.83 Å [to O(1)] are too long as to allow any charge transfer to the framework, and there is no direct coordination to any sodium ion. This site is possibly stabilised by the dipole moment of the molecule interacting with the charge distribution around the window, which depends on the local arrangement of the aluminium atoms; in a diffraction study, which yields a structure that is an average over time of all the unit cells of the sample, six-fold disorder is apparent. More detailed analysis, and experiments carried out at higher coverages, suggest that the site adjacent to the SII sodium ion also becomes occupied, but with a high degree of orientational disorder for the pyridine molecules [67].

In potassium gallozeolite-L [68], the nitrogen atom is coordinated, as one might expect, to one of the K^+ ions, K(4), at 3.00 Å distance, and is oriented so as to lie close to the channel wall, thus optimising its interactions with the framework. The position of pyridine was also investigated by energy minimisation calculations, using atom-atom potentials that included van der Waals and electrostatic terms between the molecules and the zeolite. Excellent agreement is reported between the experimentally determined and calculated locations.

4.3 Xylenes and Other Aromatics in Zeolites X and Y

Xylenes in FAU Structures. The location of the three xylene isomers in Yb, Na-Y zeolite was studied at 5 K by Czjzek et al. [69, 70]. This zeolite was chosen because the activity of Y-zeolites for reactions like isomerisation and disproportionation of xylenes increases when sodium ions are replaced by rare earth cations. For neutron diffraction, ytterbium is a good candidate because it has a large coherent cross section and relatively small incoherent and absorption cross sections compared to the other rare earth metals. The Yb^{3+} ions are not in direct contact with xylenes since they are mainly located at the position SI′ in the sodalite cage. They do however interact with the framework leading to a contraction of the lattice (the cell parameter is 24.58 Å in Yb, Na-Y compared with 24.85 Å for Na-Y [22]). Upon xylene adsorption, a further slight modification of the unit-cell parameter is observed [70], depending on the sample composition. This flexibility of the framework is also seen on adsorption of benzene. For the three isomers, two crystallographic sites A and B are found (Fig. 8). In both sites, xylene molecules are perpendicular to the [111] axis and interacting with Na^+ ions in site II. Position B is rotated by 30° with respect to position A, so that the carbon atoms lie on the intersecting mirror planes. Six equivalent orientations are found for *o*-xylene on each site by rotating the molecule through 60°. For *p*-xylene, three orientations are possible on each site, by rotating the molecule through 60° and 120°. However, *m*-xylene molecules occupy only site B at low loading and they have only three possible orientations because the two methyl groups then have short contacts to the O(1) framework oxygens (2.82 Å). At high loadings (2.5 molecules per supercage), some *m*-xylene molecules are found on site A but the favoured site remains site B. For *p*-xylene and *o*-xylene, most of the molecules are also adsorbed at low loading on site B. Site A has a higher occupation at high loading to minimise steric hindrance between methyl groups of adjacent molecules. The distance of the aromatic ring centre to the sodium cation is 2.41 Å for *p*-xylene in Yb, Na-Y. This increases to 2.57 Å for *o*-xylene and to 2.65 Å for *m*-xylene, indicating a decrease of the interaction in the following order: *p*- > *o*- > *m*-xylene.

The rotational dynamics of *p*-xylene and *m*-xylene in Na, Yb-Y have been studied by quasielastic neutron scattering (QENS) in the temperature range 100–460 K [71]. At 100 K, only methyl groups rotate while the aromatic ring stays fixed, on a time scale of 10^{-11} s. Some methyl groups, however, are blocked due to different interactions with the framework. With increasing temperature, a uniaxial rotation of the whole molecule about the axis perpendicular to the molecule sets in and the amplitude of motion increases with temperature. Isotropic spherical rotation is never reached indicating strong interaction with the cations. For a loading close to the saturation, the molecules still rotate but the amplitude is restricted by molecule-molecule interactions.

Molecular simulations have also been carried out to study the structure, energetics, and diffusion properties of *p*-xylene in Na-Y zeolite [72]. The predicted adsorption site is located inside the supercage, in front of the SII sodium cations, with the same orientation as the main species located by neutron diffraction

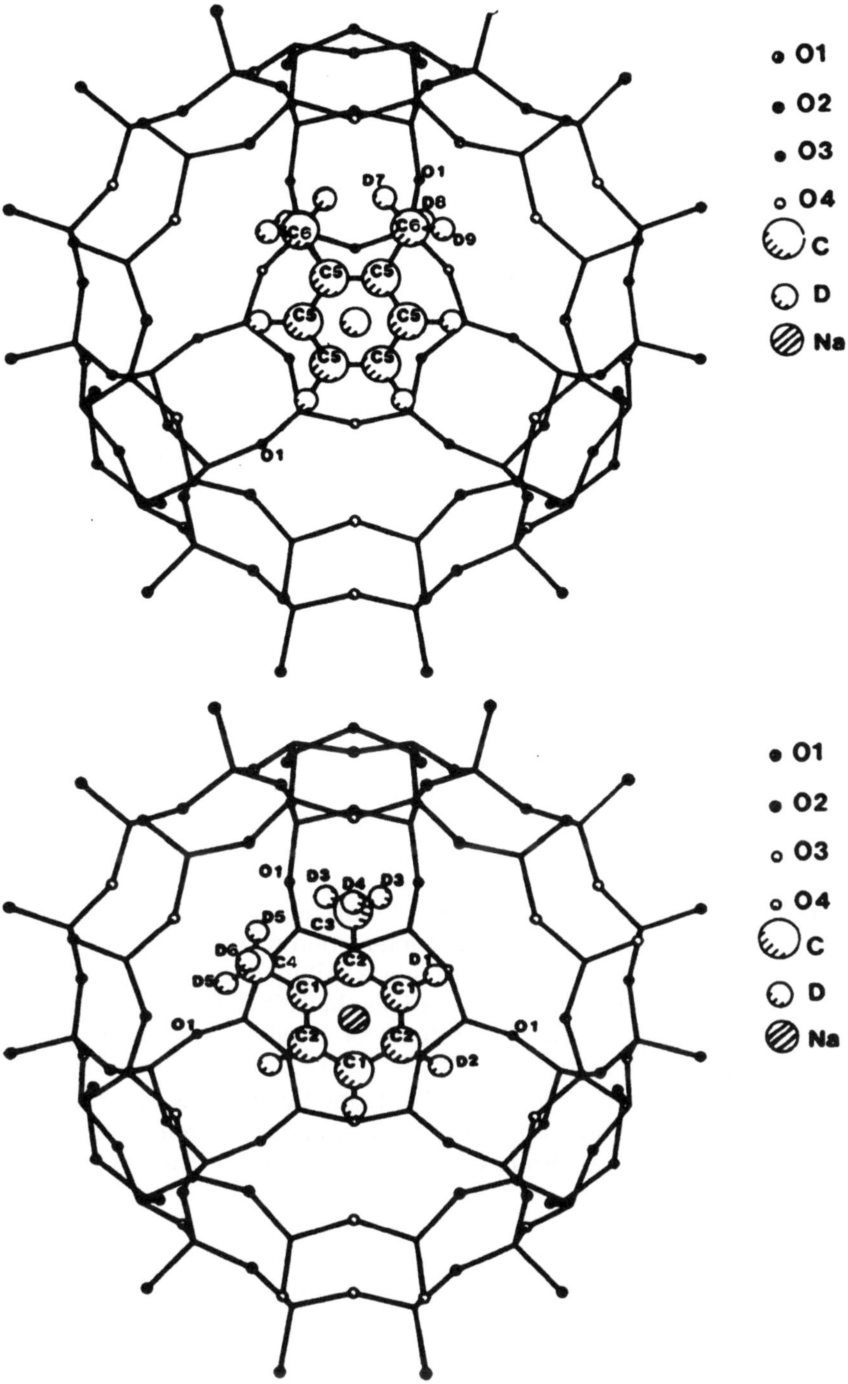

A

B

Fig. 8. Part of the unit cell of zeolite Y showing the supercage, viewed through the twelve-ring window, and the two possible sites for xylene isomers (here *ortho*) [70]

[70]. Minimisations performed at higher loadings suggest that the other orientation is due to clustering. The rotational dynamics were also investigated [72]. It was found that the rotation around the C_6 axis of the aromatic ring is much faster than the out-of-plane rotation and that this motion becomes slower as the loading increases. Furthermore, C_6 rotation was found to be fast compared to migration between adsorption sites. All these results obtained for rotational dynamics by simulations are in agreement with the experimental QENS results [71].

The diffraction studies performed on xylenes in Yb, Na-Y have been extended by Mellot et al. to Ba zeolite X [73–75]. The Ba-X zeolite is of great interest because of its higher selectivity for *p*-xylene when competitive adsorption of xylenes occurs. This zeolite is used for the industrial separation of *p*-xylene from other C_8 aromatics, (i.e. *o*- and *m*-xylene and ethylbenzene). A molecular model for the filling of the supercages by *m*-xylene and *p*-xylene has been proposed. Up to two molecules per supercage, *m*-xylene molecules are adsorbed as in Fig. 8, with the orientation B. The center of the aromatic ring is 2.44 Å from the Ba^{2+} cation in position SII. A very short distance of 2.38 Å is found between the deuterium atoms of the methyl groups and the O(1) atoms, indicative of hydrogen bond type interactions. Close to the saturation of the zeolite (loading 2.8 molecules per supercage) all *m*-xylene molecules interact with barium cations in site SII; some migration of these cations into the supercages occurs as the loading increases. However, the positions of the molecules are different from those found at low coverage. Approximately two molecules are adsorbed in position B′ and one in position A (Fig. 9). The Ba^{2+}-to-ring distance observed in position B′ (2.75 Å) is shorter than in position A (2.90 Å). The third *m*-xylene molecule provokes a reorientation of the two molecules adsorbed in site B to the less favourable B′ position in order to allow the third molecule to occupy site A. The situation was found to be more complicated after *p*-xylene adsorption. The first two molecules were found in position A (Fig. 8), at a distance of 2.8 Å from Ba^{2+}, but the third molecule could not be located. This third *p*-xylene molecule has no direct interaction with barium cations and it does not affect the two molecules in site A. Although it is less energetically adsorbed, possibly through the window, it has been found by QENS that, at high loading, the average rotational motions of *p*-xylene molecules are more blocked than those of *m*-xylene [76].

On the other hand, Descours et al. [77] found that in Na-X, *m*- and *p*-xylene are adsorbed near the SIII cation site, at low coverage, in preferance to SII. Reorientations of the molecules near SIII are observed for loadings larger than 2 molecules per supercage.

Other Aromatic Molecules in FAU Structures. The two adsorption sites determined for benzene in Na-Y were also observed for aniline in Yb, Na-Y [78]. For the site in front of the sodium cation SII, the molecule has a threefold orientational disorder (Na^+-to-ring distance: 2.59 Å). As for *m*-xylene in the same zeolite, these orientations involve short contacts of the ND_2 group with O(1) oxygens. For the window site, a sixfold orientational disorder is observed. The ND_2 group points in the direction of the O(4) atoms. Contrary to benzene, the

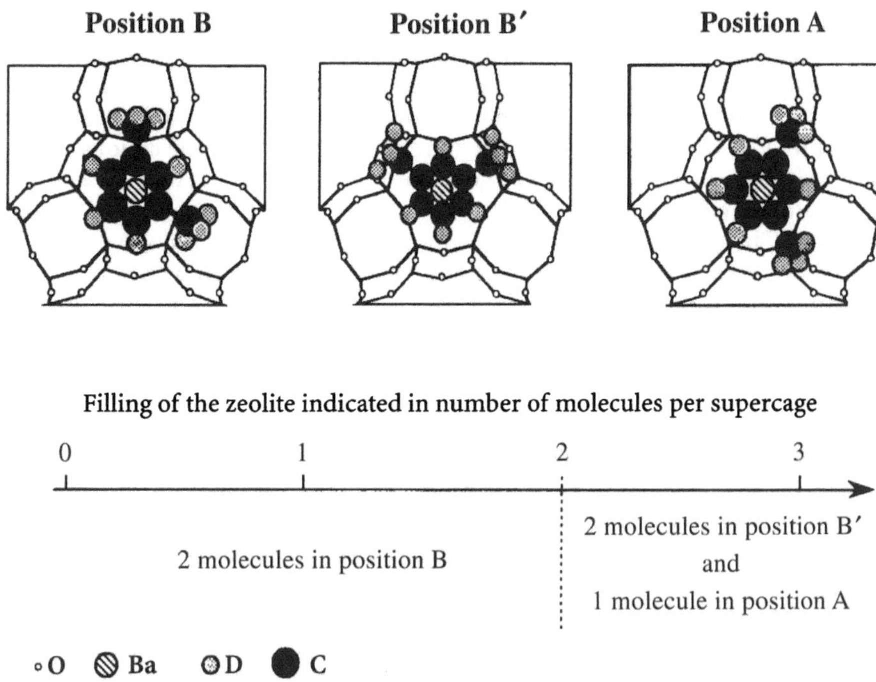

Fig. 9. Adsorption of *m*-xylene in Ba-X zeolite showing the different positions for the molecules as a function of the loading [75]

two sites are equally occupied by aniline. A contraction and a flattening of the window are observed, as in the case of benzene [22]. However, because of its size, aniline is found to be tilted out of the window plane.

Mesitylene (1,3,5-trimethylbenzene) is a bigger molecule than aniline and therefore does not fit into the twelve-ring windows. This molecule was located inside the supercage of Yb, Na-Y, with two different orientations in front of the SII sodium ions, as in Fig. 8 [79].

The preferred adsorption sites of several aromatic molecules, benzene, toluene, xylenes, mesitylene, aniline, *m*-nitroaniline, and *m*-dinitrobenzene, were investigated by molecular mechanics calculations [80]. The nonpolar molecules were all located in front of the six-ring window, with Na^+-to-ring distances in the range 2.5–2.8 Å, in good agreement with experiment. *m*-Xylene and mesitylene were found essentially in the orientation B (Fig. 8), due to attractive interaction of methyl groups to the O(1) oxygens. Aniline was located in front of the six-ring window, like the nonpolar aromatic molecules, in agreement with neutron diffraction [78]. An increase in polarity, by introducing one or two nitro substituents, leads to completely different adsorption sites. For *m*-nitroaniline, the calculation shows that the nitro group points towards the SII Na^+ cation whereas the aromatic ring crosses the twelve-ring window so that the amino group is located in the adjacent supercage. The computed structure of *m*-dinitrobenzene found the molecule in front of the four-ring window. Both structures were found to be in reasonable agreement with X-ray diffraction data.

Recently, powder neutron diffraction studies have been made on *m*-dinitrobenzene in Na-Y at different loadings, homogenisation temperature and water content [81]. The polar nitro group favours interactions with the cations rather than with the negatively charged framework. Complex behaviour is seen for these systems, with variations in the arrangements of the cations and the sorbed molecules. The most important factor controlling the adsorption is due to polar interactions; the cations take up positions where coordination by the oxygen of the framework, water or guest molecules is possible and the molecules take up positions so that the nitro groups interact with as many cations as possible. Other factors are the minimisation of steric hindrance between the framework and the guest, and sample preparation and history.

Coadsorption of aniline and *m*-dinitrobenzene in Na-Y shows two adsorption sites for each of the molecules [82]. The different molecules were distinguished by studying samples containing deuterated aniline and normal *m*-dinitrobenzene, thus exploiting the difference in the neutron scattering lengths of D and H, 6.671 and –3.739 fm, respectively. Final refinements were conducted on fully deuterated samples. All the molecules lie perpendicular to the cube diagonal of the cubic unit cell. Aniline molecules are found bound to the SII Na^+ ion and near the twelve-ring window. The dinitrobenzene molecules are far from any window at intermediate positions. Although in the diffraction experiment this arrangement appears highly disordered because of the high symmetry of zeolite Y, visible-UV spectroscopy shows a broad band around 400 nm indicating a charge transfer interaction between the two molecules, so that the molecules are arranged in a coplanar manner, stacked alternately one above the other along the unit cell diagonal. The arrangement of the sodium ions is markedly modified by the aromatic molecules and their location helps stabilise the adsorbed molecules.

4.4 Benzene in Potassium Zeolite L

Studies at room temperature and 78 K have located benzene in dehydrated zeolite L, $K_9Al_9Si_{27}O_{72}$ [83, 84]. In characterising the bare zeolite prior to adsorption of benzene the ratio of Si to Al was refined for the two crystallographically distinct framework T-sites [85]. Thus it was possible to show that the Si(1) site of the twelve-ring window is more favourably occupied by the Al atoms, in agreement with the different average T–O distances of 1.650(4) and 1.636(3) Å seen for Si(1) and Si(2), respectively. In a previous X-ray study [86], no direct refinement of the preferred occupancy of Al was possible, and both T-sites were given the average occupancy based on the chemical composition. By means of structural refinement, Fourier calculations and interactive molecular modelling, various models for the location of benzene were assessed. Although favourable in the modelling calculations, the site in the twelve-ring window was found to have zero occupancy at the coverage investigated, i.e. 1 molecule per unit cell. The sole site for benzene was found capping one of the five crystallographically distinct potassium ions in the channel side walls. The interaction between π density of the benzene molecules and the cations appears similar to that between benzene and sodium SII in Na-Y. This arrangement affords facile reorientation in the molecular plane, as indicated by deuterium NMR measurements above 150 K.

4.5 Benzene in MFI Structure

The location of eight molecules of benzene per unit cell adsorbed into ZSM-5 at 77 K has been investigated by Taylor [87, 88]. The arrangement of molecules is complicated, with three general locations. Two of these are in the intersection of the straight and sinusoidal channels, and the third is at the centre of symmetry in the straight channel. There is considerable disorder in these positions, and six independent molecules were required to fit the powder diffraction pattern. Sacerdote-Peronnet and Mentzen [89] have investigated benzene at a lower coverage, four molecules per unit cell, and saw molecules adsorbed only in the intersection. Two benzene orientations are required to fit the data. Benzene in ZSM-5 has also been studied by powder X-ray diffraction, again with indications of complex behaviour for the molecules [90].

Recently, a new study of this system has been made using high-resolution powder neutron diffraction and X-ray diffraction data collected using synchrotron radiation [91]. The structure is refined to both data sets simultaneously, thus increasing the amount of information available to define the structure. The structural model must fit well to both data sets, thus decreasing its vulnerability to systematic errors, either in the data collection or in the data analysis. The different relative scattering power of the atoms for neutrons and X-rays means that different atoms contribute differently to the two patterns, e.g. the hydrogen atoms contribute very little to the X-ray pattern, but are strong scatterers of neutrons. As well as improving the reliability of the final crystallographic model, this approach, which has already been mentioned in the section describing studies of the frameworks of zeolites, allows the exploitation of the natural complementarity of the different radiation sources.

For benzene in ZSM-5, both X-ray and neutron diffraction can be used to show the general whereabouts of the adsorbed molecules. However, because some of the chemically distinct sites are close together, and all sites are in the proximity of symmetry elements, there are superimposed positions which cannot be occupied simultaneously. Furthermore, the molecules perform significant molecular motion at room temperature, so that unravelling the details of the molecular orientations is very difficult from a single diffraction technique alone, since diffraction sees only the average scattering density. In the X-ray pattern only the positions of the carbon atoms can be observed, and the proximity of alternative sites means that different models, involving quite different orientations of the molecules, can fit the data equally well. For neutrons, both the carbon and deuterium atoms scatter strongly, and such a complex and blurred volume of scattering density is observed that it is again impossible to deduce the true orientations of the molecules. However, from a combined approach, a model that accounts for both sets of data can be developed, and this should be the model that is closest to the true positions of the adsorbed molecules.

For a coverage of 3–4 molecules per unit cell, benzene was located only in the intersection. Two crystallographic sites are required to describe the scattering density. Each of these is twofold disordered by the mirror plane that passes midway through the intersection. Hence in each intersection, one benzene molecule can be accommodated, and it may reside in any one of four actual positions. At higher coverage, 7–8 molecules per unit cell, molecules were found adsorbed into a site in the intersection, in the sinusoidal channel, and in the straight channel. Each site is twofold disordered by symmetry. The positions found agree very well with the results of theoretical predictions of the location of benzene in ZSM-5 [92, 93]. Hence it appears that the combined refinement of neutron and high-resolution X-ray data is a powerful way to unravel complex problems in the characterisation of zeolite and zeolite-adsorbate structures.

5
Location of Small Physisorbed Molecules

Attempts to induce molecular hydrogen to form a super-fluid Bose-condensed phase have tried to suppress the temperature of the bulk liquid-solid transition at 13.8 K, by confining the hydrogen to small volumes in porous media. NMR studies of H_2 adsorbed in zeolite 13X appeared to show a suppression of the transition to around 8 K. However, time-of-flight powder neutron diffraction measurements of D_2 in zeolite 13X, which is expected to behave in a manner similar to that of H_2, showed that the deuterium molecules are adsorbed at two distinct sites in the host [94]. Hence the deuterium becomes an integral part of the crystal structure, and can no longer be viewed in terms of a bulk solid or liquid phase. No details of the sites were given. However they differ in stability, the second site starting to fill once the more stable site becomes saturated. When the second site is full, a bulk phase forms on the surface of the zeolite.

The physisorption of rare gases and small molecules in MFI and AFI structures [13] has been studied by Llewellyn et al. [96, 97] and by Coulomb et al. [97]. Stepped isotherms were measured for these systems and interpreted by localised adsorption. In situ neutron diffraction experiments were performed at different points of the adsorption isotherm, at low temperature. In silicalite I, it was found that argon and krypton undergo a liquid-solid-like phase transition at high loadings but that methane remained disordered [96]. Despite the difference in size between Ar and Kr, the uptakes at saturation are the same. The adsorption sites in silicalite I appear homogeneous to physisorption, which means that these atoms do not seem to distinguish between the straight or sinusoidal channels and the intersections. The solid-like phases of Ar and Kr are identical in structure and commensurate with the silicalite I microporous network. Similar results were obtained for nitrogen and carbon monoxide in the MFI structure, except that two substeps were found in the isotherms of both molecules [96].

The first one was assigned to a phase transition from a disordered ("fluid") phase to a lattice fluid-like phase and the second one to a crystalline-like solid phase.

A more quantitative analysis was carried out for H_2, Ar, CH_4 and CF_4 adsorbed into $AlPO_4$-5 and $SAPO_4$-5 [97]. The structure of this aluminophospate is much simpler than the MFI structure and consists of one-dimensional 12-membered ring channels. The molecules appear to be extensively disordered in the one-dimensional channels of the zeolite. For methane, a transition is reported to occur during loading from a disordered fluid phase to a more-ordered arrangement of molecules. A model comprising chains of CD_4 "dimers" (four molecules per unit cell) gives good agreement with the measured powder diffraction pattern.

In these systems, neutron diffraction has been used to obtain information on short- and long-range order. However, no Rietveld refinement has been attempted so far to derive the positions of the physisorbed molecules.

6 Single-Crystal Studies on Natural Hydrated Minerals

Neutron diffraction has been used to locate the water molecules and to study hydrogen bonding in several natural zeolites. The heavier atoms had usually been previously located in X-ray structure refinements but neutron diffraction yielded more precise locations for the hydrogen atoms. The Si/Al order or disorder can also be evaluated more precisely from neutron than from X-ray diffraction because of the larger difference in the scattering amplitudes of these elements.

It appears that the first neutron diffraction study to locate water in zeolite single crystals was performed by Torrie et al. [98] in natrolite (*vide infra*). This was also the first water/zeolite system to be studied by inelastic neutron scattering [99]. Other pioneering studies were made by Bartl in leonhardite [100] and by Ferraris et al. in analcime [19]. Analcime, $Na_{16}Al_{16}Si_{32}O_{96} \cdot 16\,H_2O$, has Si and Al atoms randomly distributed and the 16 Na cations have a disordered distribution over the 24 sites available to them. When there is such a disorder, the atoms are more difficult to locate precisely, especially the water molecules, and large temperature factors are usually obtained.

With natrolite, $Na_2Al_2Si_3O_{10} \cdot 2H_2O$, in addition to the work of Torrie et al. [98], two more recent neutron refinements have been performed by Pechar et al. at room temperature [101], and by Artioli et al. at 20 K [102]. The natural single crystals used in both studies possessed fully ordered Si/Al distributions. This could be checked from the refinement of the Al and Si scattering lengths and also from the large differences between the lengths of the *b* and *a* unit-cell parameters ($b - a$) a criterion introduced by Alberti and Vezzalini [103]. The final atomic coordinates are similar in both neutron studies, with lower estimated standard deviations in the measurement performed at 20 K. In particular the isotropic B factor for the water molecules is smaller by a factor of 2 at 20 K (the B factors for the hydrogens are 1.87 and 2.1 $Å^2$). The mean Si–O distances are

1.620 and 1.622 Å and the mean Al–O bond length is 1.747 Å. The sodium cation is coordinated to 2 water oxygens (distance 2.37 and 2.39 Å) and to 4 framework oxygens (two short distances of 2.37 and 2.39 Å and two longer ones at 2.5 and 2.6 Å). The water molecules are all equivalent, each water oxygen is linked to 2 sodium ions (distances given above), and the hydrogens are bonded asymmetrically to 2 oxygens of the framework (distances 1.88 and 2.10 Å). Similar positions were obtained by X-ray diffraction at room temperature [104]. However for the water molecules the Ow–H distances from the X-ray study are too short (mean value of 0.66 Å), the hydrogen bonds with the framework are too long (2.21 and 2.35 Å) and the H–O–H angle is too large: 117° compared with 108° in [98], 107.9° in [102] and 110.3° in [101] (in the gas phase the value is 104.5°). It should be noted that in the X-ray study [104], some dehydration of the crystal was observed after a long period of data collection, and this was attributed to absorption of radiation energy.

The structure of another fibrous zeolite, edingtonite, $Ba_2Al_4Si_6O_{20} \cdot 7H_2O$, has been studied by neutron diffraction at 294 K [105]. The ideal unit cell has 8 water molecules but some dehydration occurs at room temperature. Complete Si/Al ordering was also found in this zeolite. Each barium ion lies in the centre of the channel which is pulled into an elliptical shape by bonding to 6 framework oxygens (at 2.89–3.04 Å) and to 4 water oxygens. Four water molecules are coordinated to the barium cation and linked to oxygens of the framework. Weak interaction of water with the framework is reflected by the large distances between the hydrogens and the oxygen atoms (1.915–2.103 Å). The crystal structure of another sample was studied at 293 and 100 K by Belitsky et al. [106], and refined parameters in agreement with those reported by Kvick and Smith [105] were obtained. The neutron data gave essentially the same results for both temperatures, but with a lower symmetry at 100 K due to shifts in one hydrogen position. In addition to the neutron diffraction work, a combined ^{27}Al, ^{1}H and ^{2}H study gave evidence of a structural phase transition below 200 K and of anomalous librational motions for some water molecules [106].

The structure of scolecite, $CaAl_2Si_3O_{10} \cdot 3H_2O$, is similar to that of natrolite but the cations and water molecules occupy different positions. There is also a small monoclinic distortion and the structure has been described in different space groups and unit cells. Neutron diffraction studies have been performed at room temperature [107, 108] and at 20 K [109]. The calcium ion is strongly coordinated to 3 water oxygens and linked to 4 oxygens of the framework (at 2.48–2.59 Å). The 3 water molecules are shown in Fig. 10. All water molecules are involved in one strong and one weak hydrogen bond to the framework (due to Al/Si ordering). The H–O–H angle varies from 106.6° to 111.1°.

Heulandite has an ideal composition, $Ca_4Al_8Si_{28}O_{72} \cdot 24H_2O$. If severely dehydrated, there is an irreversible collapse of the lattice and of the zeolite channels. Hambley and Taylor have studied at room temperature the hydrated and partially dehydrated forms of a single crystal of heulandite close to the ideal composition [110]. It was found that dehydration leads to significant changes in the lattice constants, to a depletion in the water and cation sites and to the occupation of a new water and cation site. The influence of this cation and water movement upon dehydration was discussed in the context of the thermal stability of

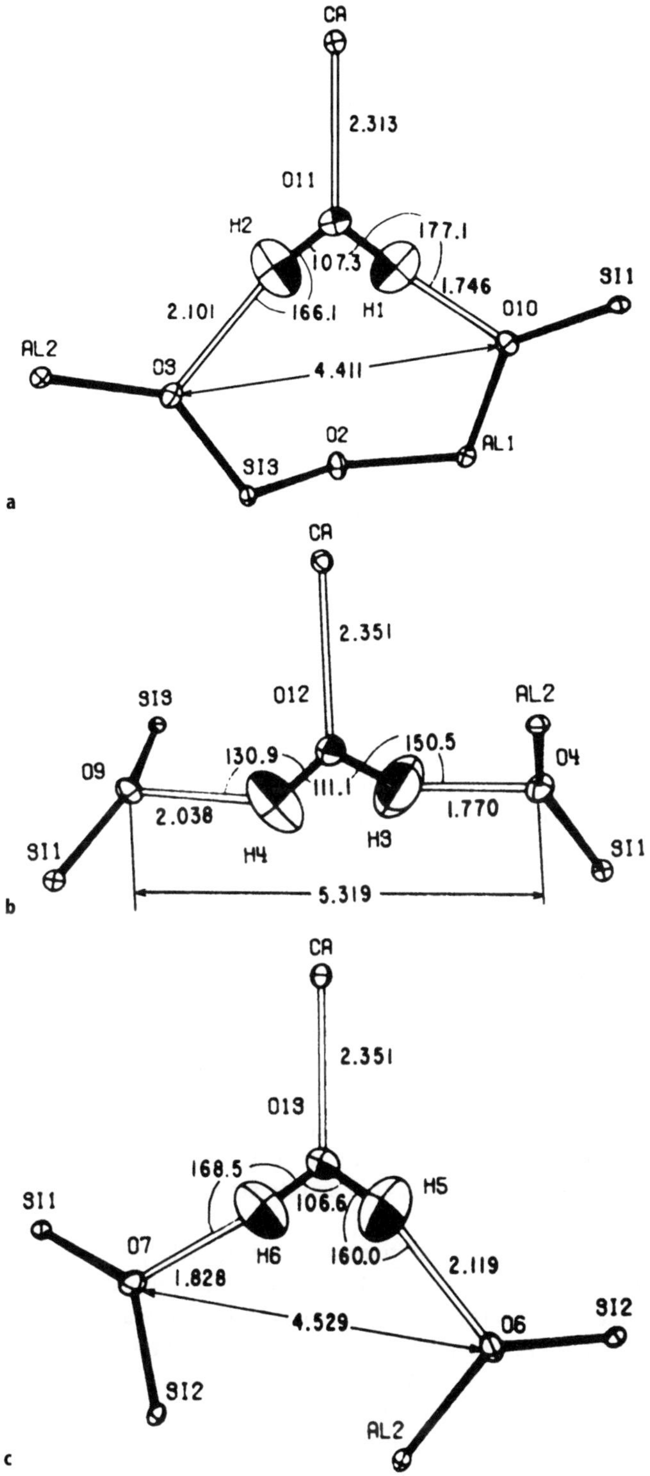

Fig. 10. The hydrogen bond scheme in scoletite at 20 K [109]

this zeolite. In clinoptilotite, which has a higher Si/Al ratio and where much of the calcium is replaced by monovalent cations (Na and K), the occupancy of additional sites was suggested as the reason for the greater thermal stability of this structure.

The bonding of water molecules in another fibrous zeolite, thomsonite, of composition close to $NaCa_2Al_5Si_5O_{20} \cdot 6H_2O$, was studied by Pluth et al. at 293 K [111]. Again, precise positions for all the heavier atoms and correct but imprecise positions for the hydrogen atoms had previously been derived from X-ray diffraction [112]. The neutron refinement gave precise positions for all atoms, including the hydrogen. The refinement indicated Si/Al ordering. There are 4 independent water molecules. Each water oxygen is bonded either to one Ca, or to two Na or Ca ions that together form a site of mixed occupancy. Two framework oxygens, O1 and O2, are bonded to two hydrogens, while two other oxygens, O3 and O4, are only bonded to one hydrogen. There is some positional disorder combined with thermal disorder which is reflected by the large isotropic temperature factors for some water atoms.

As in analcime, the Si/Al atoms in brewsterite, $(Sr, Ba)_2[Al_4Si_{12}O_{32}] \cdot 10H_2O$, are randomly distributed. A neutron diffraction study on a natural single crystal with little Ba was carried out by Artioli et al. at room temperature [113]. Multiple proton positions were found for 3 out of the 4 types of water molecules, and they could not be related to a local arrangement of Si and Al atoms. The Al substitution was estimated from the T–O distances. Multiple choices were derived for short-range order models. It was concluded that the bonding scheme was tolerant of a variety of Al distributions, and that each bond distance or angle was subject to a minor perturbation in response to the actual distribution of nearby atoms.

The disordered water distribution in a sample of stellerite, of composition $Ca_{7.7}Na_{0.3}(Al, Si)_{72}O_{144} \cdot 50H_2O$, was studied at room temperature by Miller and Taylor [114]. Stellerite, stilbite and barrerite have the same aluminosilicate framework structure but vary in the concentrations of sodium and calcium cations. Stellerite has the highest symmetry structure and the single crystal neutron diffraction study was described in the space group *Fmmm*. Contrary to previous examples, the calcium ion in stellerite is completely isolated from the framework by an encompassing sphere of disordered water molecules. The hydrogen atoms point away from calcium and interact weakly with framework oxygen atoms. Attempts were made to study dehydration. However, the crystals disintegrated when heated to 100 °C in air or in vacuum, indicating that water molecules are an essential part of this zeolite structure, even if weak hydrogen bonding is observed.

A neutron diffraction study of gismondine, $Ca_{3.91}Al_{7.77}Si_{8.22}O_{32} \cdot 17.6H_2O$, was carried out at 15 K by Artioli et al. [115]. The atomic coordinates of all framework atoms, calcium, and the first 4 water oxygens was in agreement with a previous X-ray refinement performed at room temperature [116]. An essentially ordered Si/Al distribution was found in the neutron study. Again, neutrons are essential to describe precisely the hydrogen bonding system. As shown in Fig. 11, the cation is attached to one side of the eight-ring channels, coordinating 2 framework oxygens (O4, O8) and 4.3 water molecules. Two possible configurations are envisaged: 6-coordination of calcium with 70% occurrence (O4, O8,

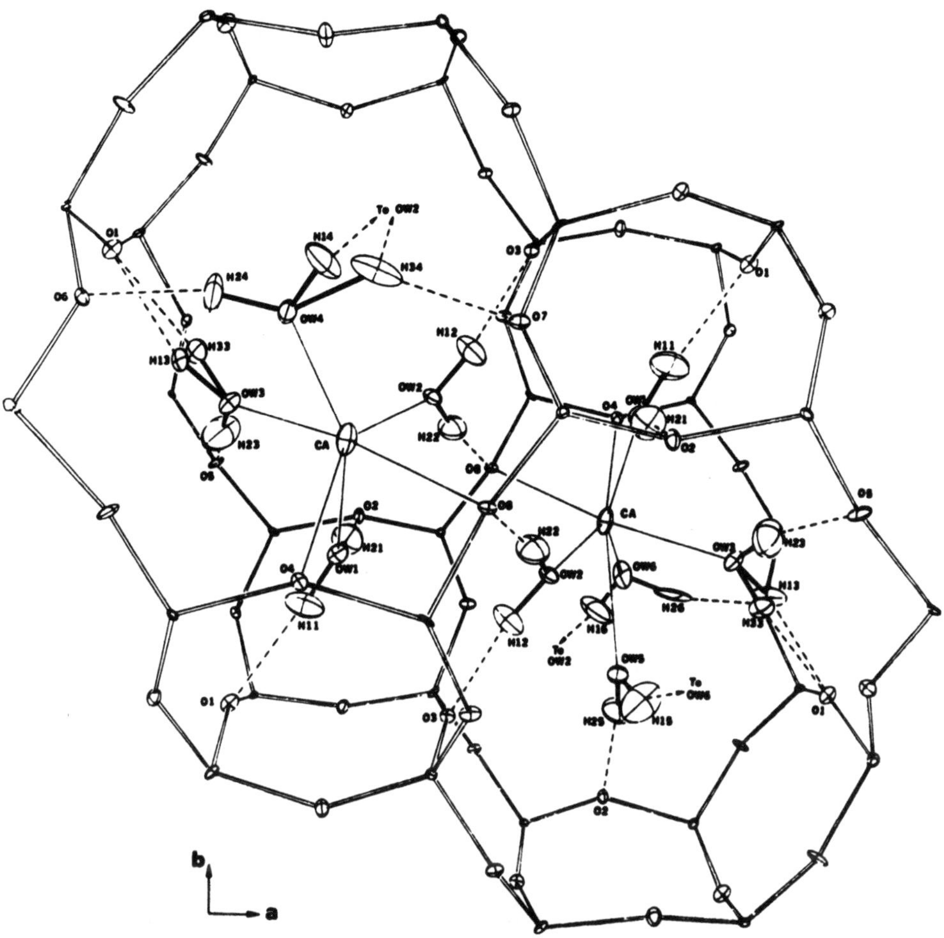

Fig. 11. Cation coordination and hydrogen bonding in gismondine. The Ca^{2+} ion on the left of the picture is shown six coordinated (70% occurrence) and that on the right is shown seven coordinated (30% occurrence) [115]

OW1, OW2, OW3, OW4), and 7-coordination with 30% occurrence (O4, O8, OW1, OW2, OW3, OW5, OW6). In the latter case, the 2 sites OW5 and OW6 replace OW4. Apparently, O8 is the only framework oxygen found in a zeolite involved in a Ca interaction and H-bonding at the same time. Proton positions were determined for all water molecules, disordered hydrogen sites were found for OW3 and OW4. No simple relationship could be proposed between the hydrogen bonding system and the dehydration process.

The bonding of water molecules in yugawaralite, $CaAl_2Si_6O_{16} \cdot 4H_2O$, is even more complex than in gismondine. Although yugawaralite has an ordered Si/Al distribution, 7 different water oxygen locations with only 2 fully occupied oxygen sites and a multitude of alternative hydrogen sites were found by neutron diffraction by Kvick et al. [117]. The calcium cation is bonded to 4 framework oxygens (at 2.48–2.558 Å) and to 4 water oxygens (at 2.315–2.541 Å). Hydrogen bond distances were found in the range 1.87 to 2.71 Å, indicating a wide range of interaction strength. A water molecule in a partially occupied site was found to be not coordinated to calcium and would therefore be eliminated first upon heating.

The crystal structure of bikitaite, $Li_2Al_2Si_4O_{12} \cdot 2H_2O$, was studied at 13 and 295 K by Ståhl et al. [118]. The structures obtained at the two temperatures are very similar, except for thermal parameters (the B factor averaged over all H-atoms decreases from 5.7 Å^2 at 295 K to 2.33 Å^2 at 13 K). The structure is triclinic, space group P1, with a high degree of Si/Al ordering. The lithium cations are each coordinated to 3 framework oxygens and 1 water oxygen. The highlight of this system is that the two independent water molecules are linked to each other and form one-dimensional chains, stabilized by coordination to the cations (Fig. 12). Such chains of hydrogen-bonded water molecules were found in VPI-5, here a triple helix, by synchrotron powder diffraction [39]. Two hydrogen atoms, H11 and H21, are not involved in hydrogen bonding since there are no contacts shorter than 2.54 Å (Fig. 12). Furthermore, hydrogen bond distances are 1.997 and 2.002 Å at 295 K, indicating weak bonding. This suggests

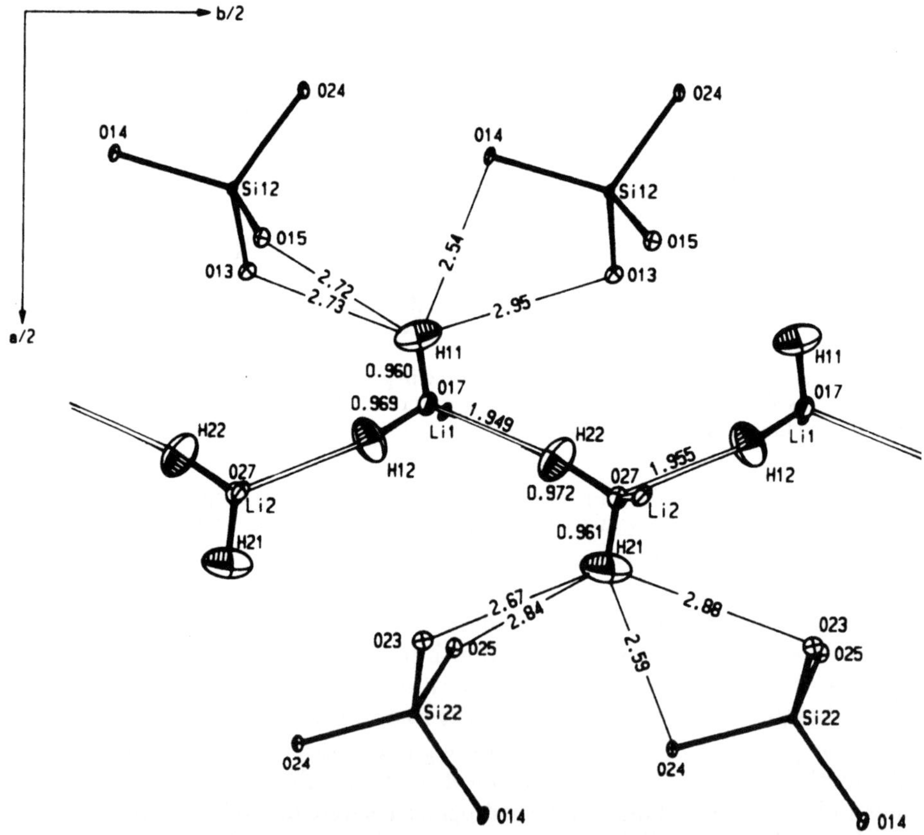

Fig. 12. The water chain in bikitaite at 13 K. Hydrogen–oxygen distances less than 3 Å are given [118]

a large mobility of the water molecules, or even proton conduction by transfer of the protons H12 and H22 within the bonds followed by water reorientation. However, ^{1}H NMR spin-relaxation studies identified only one type of dynamic process in the temperature range 224–418 K, that is 180° flip motions of the water molecule about its quasi twofold axis [119].

Laumontite, ideally $Ca_2Al_4Si_8O_{24} \cdot 16H_2O$, dehydrates under normal conditions (295 K, 1 atm) to leonhardite with 14 H_2O [100]. A complete rehydration process appears to be difficult. Artioli et al. have studied two samples at 15 K by neutron diffraction: a natural leonhardite and a specimen soaked in water in an attempt to obtain laumontite [120]. Structure refinements proved however both crystals to be partially hydrated with 13.4 and 14.2 water molecules per unit cell, respectively. At the lowest water loading, 8 positions for water oxygens were located and 11 positions at the highest loading. This zeolite is the one showing by far the highest degree of disorder for water. This shows that an ordered Si/Al distribution does not necessarily yield an ordered water structure. The calcium cation is linked to 4 framework oxygens and to water molecules. The coordination of calcium varies between 6 and 7. The presence of 6 and 7 coordinated calcium ions was noted in gismondine (*vide supra*). Some water molecules are not bonded to calcium. They take part in a complex hydrogen bonding net across the channel, linking water molecules connected to different calcium ions.

Some conclusions can be drawn from these neutron diffraction studies. The interaction of the cation with the water oxygen is usually strong and is essential for the location of the water molecule. The interaction of water hydrogens with the oxygens of the framework is weaker. A compilation was made by Kvick for natural zeolites [121]. In the majority of cases, the water molecules are hydrogen bonded to the framework as well as coordinated to the cations. The hydrogen bonds tend to be asymmetric, the shortest distance being 1.764 Å (Fig. 10a). There is a preference for oxygen atoms bonded to 1 Si and 1 Al to be the hydrogen bond acceptors (Fig. 10c being the only exception). Weakly bonded water has an H–O–H angle close to the gas phase value (104.5°) with O–H distances only slightly longer than the value of 0.9572 Å observed in the gas phase. In the compilation made for 27 ordered water molecules [121], the H–O–H angle varies between 96.9° and 112.5°. The general trend correlating the lengthening of the O–H distance with increasing hydrogen bond strength is confirmed in the natural zeolites.

7
Proton Positions and Hydronium Species

Location of protons in zeolites is of interest because they constitute potential Brønsted acid sites. Proton positions can be determined more easily with neutron diffraction as compared to X-ray diffraction studies owing to their enhanced scattering power. Quite often, the *protonated* form of the zeolite is used, despite the background due to the incoherent scattering, to simplify the preparation of the samples: during dehydration, zeolites containing hydrogen ions may suffer significant framework degradation. The first powder neutron diffraction experiments were performed by Jirak et al. for a series of Na, H-Y samples with different degrees of cation exchange [122–124]. The signal-to-noise ratio was rather poor and the pattern was limited up to 2θ values of approximately 45°. Nearly all the hydrogen atoms were found in the H3 and H1 positions. The hydrogen atoms are coplanar with the T-O-T atoms, the protons lying on the symmetry axis. Protons in H1 and in H3 are oriented into the supercage and into the sodalite cage, respectively. Distances between O and H atoms were calculated in the range 0.74–1.26 Å. However, all the atomic parameters for the protons and the cations were not refined and large estimated standard deviations were reported.

In La-Y, two different proton positions were found by Cheetham et al. [125]. In this study, protons attached to framework oxygen O4 were localised at a distance of 1.2 Å. This observation of H4, also pointing into the supercage, is unusual because in all other studies occupation near O4 is zero or close to zero [122 - 124, 126, 127]. Furthermore, the hydrogen of a hydroxyl group attached to the La^{3+} ion was clearly seen, indicating that hydrolysis of the cation in the cavity occurred during dehydration.

In D-RHO, a bridging hydroxyl group has been localized by Fischer et al. [29]. The position of the deuterium atom was refined, leading to an O–D distance of 0.96 Å. As in previous work [122–124], the arrangement of the T-atoms and the H-atom around the bridging O-atom is approximately planar.

The determination of hydrogen positions in Y zeolites has been reinvestigated by Czjzek et al. [126]. In this work, both H-Y and D-Y samples were studied. Owing to their negative scattering length for neutrons (Fig. 1) protons appear as negative peaks in difference Fourier maps for H-Y samples, whereas deuterium, which has a positive scattering length, appears as positive peaks for D-Y samples. At high proton content, the preferred proton site is near O1, followed by O3, then O2 (see Fig. 13). The proton located near O2 points into the six-ring window between the sodalite cage and the supercage. Contrary to previous work this site has a non-negligible occupation: approximately 10 protons per unit cell. For the H-Y sample, an average value of 0.943 Å is obtained for the O–H distance. For a D-Y sample containing 15 water molecules per unit cell, the deuterium atom near O3 was found to be shifted in the direction of a water molecule located on the site I′ in the sodalite cage. The distance Ow–D3 is 1.16 Å while the distance O3–D3 is 2.04 Å. Hence the formation of a hydronium ion in the sodalite unit was suggested.

Another study on D-Y was made by Sun and Seff using pulsed neutron powder diffraction [127]. Deuterium atoms were only located near O1 and O3, with a rather long O–D distance: 1.26 Å. Approximately 8 deuterium atoms per unit cell could not be located.

The distribution of acid sites in H-SAPO-37, the silicoaluminophosphate analogue of zeolite Y, was studied by Bull et al. [128]. For the protons, only their occupancies were refined. The final refinement led to site occupancies following the order O1 > O2 > O3, a distribution different from H-Y [126]. Upon benzene adsorption, protons could only be found at O2 [63]. Although a proportion of protons could not be detected, it appears that some protons have moved to O2 in order to bind to the benzene located above the six-ring (*vide supra*).

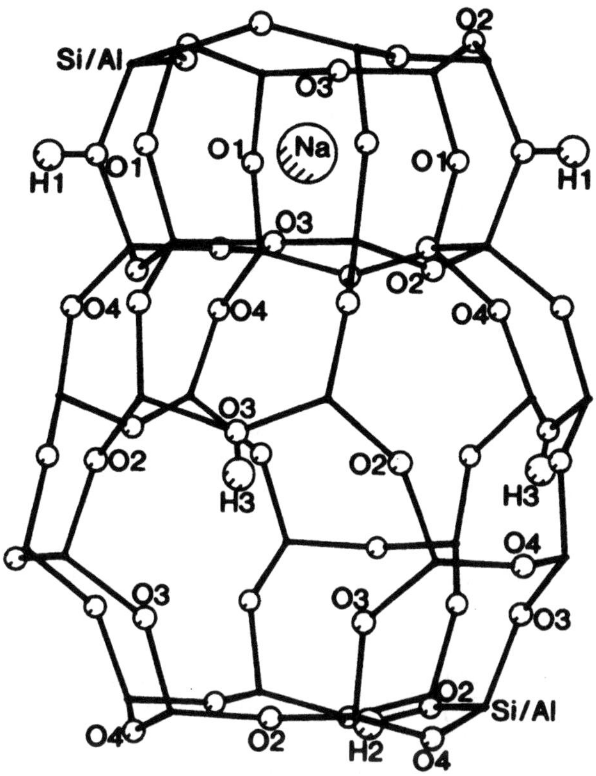

Fig. 13. Part of the structure of Na, H-Y showing a sodalite cage and a hexagonal prism, indicating proton positions near O1, O2 and O3 [126]

A much debated question is whether water can be protonated on Brønsted sites in zeolites. Conflicting interpretations of the spectroscopic results can be found. Ab initio calculations performed so far on small clusters are in favour of a water molecule attached to the acid site via two hydrogen bonds, in a neutral complex [129, 130]. Studies of water adsorbed into H-ZSM-5 by incoherent inelastic neutron scattering spectroscopy also indicated that for this system the neutral complex was the preferred mode of adsorption [131]. However, a recent powder neutron diffraction study of water in H-SAPO-34 [132] confirms the existence of hydronium ions. In this silicoaluminophosphate analogue of chabazite, two acid sites were found. Two different water molecules are associated with these sites. One is hydrogen bonded to an acid site on the six-ring, but the other is clearly a hydronium species, sitting in the eight-ring channel of the structure.

8 Concluding Remarks

When applied to large single crystals of natural zeolites, or to microcrystalline natural and synthetic powders, neutron diffraction yields detailed information on the structure and ordering of the tetrahedral atoms in the framework, the positions of cations and water molecules, and the nature of any hydrogen bonding in the channels. For dehydrated compounds, the method is effective in locating sorbed species, such as hydrocarbon and small physisorbed molecules. This information is vital for an understanding of the ion-exchange, molecular-sieving and catalytic properties of zeolites, and for the development of theoretical models of zeolite behaviour.

When carrying out neutron diffraction studies, there is frequently a lot of information already available from previous X-ray diffraction measurements. The two techniques are in many ways complementary, as both methods look at the same structure, but the atoms have different scattering power for X-rays and neutrons, so different aspects of a structure are emphasised in the data analysis. The absence of a form factor and the narrower range in the scattering powers of atoms mean that refinements from neutron data are generally considered more accurate than those from X-rays. Hence neutrons have been used to refine more detailed structural models than is possible solely from X-ray data. Where structures have been solved from powder data using high-resolution laboratory or synchrotron X-rays, it is possible to obtain the basic heavy-atom structural model from the X-ray data, and complete the determination of the light-atom structure with neutrons, as, for example, for the novel gallophosphite [48]. The final and most complete structural model is frequently refined from the neutron data alone, as this contains information about all atoms present.

Crystallographic studies using powders suffer from more problems than single-crystal studies, because peaks are obscured by the one-dimensional nature of the powder diffraction pattern. In principle, these can be disentangled and recovered in an accurate Rietveld [11] refinement, yet there are always problems and ambiguities in any structural analysis that are exacerbated using the powder technique. The greater the amount of reliable information that can be included in the analysis of a crystal structure, the more likely it is that the refined structure will be accurate, and free from the effects of systematic errors introduced in the collection of the data or in the data analysis. Chemical information in the form of bond distance and angle restraints can help the stability of a refinement, and ensure that the structural parameters have sensible values. Including additional data sets can also be an advantage. Hence for complex structural refinements, joint fitting to neutron and X-ray data sets is possible using modern Rietveld refinement programs (e.g. [133–135]), to harness the full complementarity of X-ray and neutron data, as was carried out, for example, in the studies of ferrierite [35] and benzene in ZSM-5 [91].

Whether used as a unique tool in its own right, or in conjunction with complementary X-ray studies, neutron diffraction is a powerful method of investigating crucial structural aspects of zeolites and other microporous materials.

References

1. Newsam JM (1986) Physica 136B:213
2. Baerlocher Ch (1986) Zeolites 6:325
3. Baerlocher Ch, McCusker LB (1994) In: Jansen JC, Stöcker M, Karge HG, Weitkamp J (eds) Advanced zeolite science and applications, Lectures of the 10th Int. Zeolite Conf. Summer School, Wildbad Kreuth, Germany, July 14-16, 1994. Elsevier, Amsterdam, 1994; Stud Surf Sci Catal 85:391
4. McCusker LB (1994) In: Weitkamp J, Karge HG, Pfeifer H, Hölderich W (eds) Zeolites and related microporous materials: State of the art 1994, Proc. 10th Int. Zeolite Conf., Garmisch-Partenkirchen, Germany, July 17-22, 1994. Elsevier, Amsterdam, Stud Surf Sci Catal 84:341
5. McCusker LB (1991) Acta Cryst A47:297
6. Cheetham AK, Wilkinson AP (1992) Angew Chem Int Ed Engl 31:1557
7. Bacon GE (1975) Neutron diffraction, 3rd ed. Clarenden Press, Oxford
8. Squires GL (1978) Introduction to the theory of thermal neutron scattering. Cambridge University Press, Cambridge
9. Jobic H (1992) Spectrochim Acta 48A:293
10. Wilson AJC (ed) (1992) International tables for crystallography, vol C. Kluwer Academic Publishers, Dordrecht
11. Rietveld HM (1969) J Appl Cryst 2:65
12. Fitch AN (1986) In: Freer R, Dennis PF (eds) Kinetics and mass transport in silicate and oxide systems. Materials Science Forum 7:45
13. Meier WM, Olson DH, Baerlocher Ch (1996) Zeolites 17:1
14. Newsam JM (1986) J Chem Soc Chem Commun 1295
15. Newsam JM (1988) J Phys Chem 92:445
16. Loewenstein W (1954) Am Mineral 39:92
17. Yang J, Xie D, Yelon WB, Newsam JM (1988) J Phys Chem 92:3586
18. Yelon WB, Xie D, Newsam JM, Dunn J (1990) Zeolites 10:553
19. Ferraris G, Jones DW, Yerkess J (1972) Krist 135:240
20. Newsam JM, Jacobson AJ, Vaughan DEW (1986) J Phys Chem 90:6858
21. Mortier WJ (1982) Compilation of extra framework sites in zeolites. Butterworths, London
22. Fitch AN, Jobic H, Renouprez A (1986) J Phys Chem 90:1311
23. Forano C, Slade RCT, Krogh Andersen E, Krogh Andersen IG, Prince E (1989) J Solid State Chem 82:95
24. Hriljac JA, Eddy MM, Cheetham AK, Donohue JA, Ray GJ (1993) J Solid State Chem 106:66
25. Eddy MM, Cheetham AK, David WIF (1986) Zeolites 6:449
26. Fischer RX, Bauer WH, Shannon RD, Staley RH, Vega AJ, Abrams L, Prince E (1986) Zeolites 6:378
27. Fischer RX, Bauer WH, Shannon RD, Staley RH, Vega AJ, Abrams L, Prince E (1986) J Phys Chem 90:4414
28. Bauer WH, Fischer RX, Shannon RD, Staley RH, Vega AJ, Abrams L, Corbin DR, Jorgensen JD (1987) Z Krist 179:281
29. Fischer RX, Baur WH, Shannon RD, Staley RH (1987) J Phys Chem 91:2227
30. Fischer RX, Baur WH, Shannon RD, Staley RH, Abrams L, Vega AJ, Jorgensen JD (1988) Acta Cryst B44:321
31. Corbin DR, Abrams L, Jones GA, Eddy MM, Harrison WTA, Stucky GD, Cox DE (1990) J Am Chem Soc 112:4821
32. Corbin DR, Abrams L, Jones GA, Eddy MM, Stucky GD, Cox DE (1989) J Chem Soc Chem Commun 42
33. Parise JB, Corbin DR, Gier TE, Harlow RL, Abrams L, Von Dreele RB (1992) Zeolites 12:360
34. Parise JB, Corbin DR, Abrams L, Northrup P, Rakovan J, Nenoff TM, Stucky GD (1994) Zeolites 14:25

35. Morris RE, Weigel SJ, Henson NJ, Bull LM, Janicke MT, Chmelka BF, Cheetham AK (1994) J Am Chem Soc 116:11849
36. Bennett JM, Richardson Jr JW, Pluth JJ, Smith JV (1987) Zeolites 7:160
37. Richardson Jr JW, Pluth JJ, Smith JV (1988) Acta Cryst B44:367
38. Richardson Jr JW, Smith JV, Pluth JJ (1989) J Phys Chem 93:8212
39. McCusker LB, Baerlocher Ch, Jahn E, Bülow M (1991) Zeolites 11:308
40. Richardson JW, Vogt ETC (1992) Zeolites 12:13
41. Bennet JM, Cohen JP, Flanigen EM, Pluth JJ, Smith JV (1983) Am Chem Soc Symp Ser 218:109
42. Qiu S, Pang W, Kessler H, Guth JL (1989) Zeolites 8:440
43. Richardson JW, Pluth JJ, Smith JV (1987) Acta Cryst C43:1469
44. Mora AJ, Fitch AN, Cole M, Goyal R, Jones RH, Jobic H, Carr SW (1996) J Mater Chem 6:1831
45. Harrison WTA, Gier TE, Nicol JM, Stucky GD (1995) J Solid State Chem 114:249
46. Nenoff TM, Harrison WTA, Gier TE, Nicol JM, Stucky GD (1992) Zeolites 12:770
47. Nenoff TM, Harrison WTA, Stucky GD, Nicol JM, Newsam JM (1993) Zeolites 13:506
48. Morris RE, Harrison WTA, Nicol JM, Wilkinson AP, Cheetham AK (1992) Nature 359:519
49. Roberts MA, Fitch AN (1991) J Phys Chem Solids 52:1209
50. Fitch AN, Jobic H, Renouprez A (1985) J Chem Soc Chem Commun 284
51. Fitch AN (1986) In: Catlow CRA (ed) High resolution powder diffraction. Materials Science Forum 9:113
52. Eulenberger GR, Shoemaker DP, Keil JG (1967) J Phys Chem 71:1812
53. Renouprez A, Jobic H, Oberthur RC (1985) Zeolites 5:222
54. Demontis P, Yashonath S, Klein ML (1989) J Phys Chem 93:5016
55. Uytterhoeven L, Dompas D, Mortier WJ (1992) J Chem Soc Faraday Trans 88:2753
56. Zibrowius B, Caro J, Pfeifer H (1988) J Chem Soc Faraday Trans 1 84:2347
57. Liu S-B, Ma L-J, Lin M-W, Wu J-F, Chen T-L (1992) J Phys Chem 96:8120
58. Wu J-F, Chen T-L, Ma L-J, Lin M-W, Liu S-B (1992) Zeolites 12:86
59. Pearson JG, Chmelka BF, Shykind DN, Pines A (1992) J Phys Chem 96:8517
60. Auerbach SM, Bull LM, Henson NJ, Metiu HI, Cheetham AK (1996) J Phys Chem 100:5923
61. de Mallmann A, Barthomeuf D (1996) In: Murakami Y, Iijima A, Ward JW (eds) New developments in zeolite science and technology, Proc. 7th Int. Zeolite Conf. Tokyo, Japan, August 17–22, 1986. Kodansha, Tokyo and Elsevier, Amsterdam, p 609
62. Lechert H, Wittern KP (1978) Ber Bunsenges Phys Chem 82:1054
63. Bull LM, Cheetham AK, Powell BM, Ripmeester JA, Ratcliffe CI (1995) J Am Chem Soc 117:4328
64. Eberly PE (1968) J Phys Chem 72:1042
65. Michel D, Germanus A, Pfeifer H (1982) J Chem Soc Faraday Trans 1 78:237
66. Goyal R, Fitch AN, Jobic H (1990) J Chem Soc Chem Commun 1152
67. Goyal R (1993) PhD thesis, Keele University, UK
68. Wright PA, Thomas JM, Cheetham AK, Nowak AK (1985) Nature 318:611
69. Czjzek M, Vogt T, Fuess H (1989) Angew Chem Int Ed Engl 28:770
70. Czjzek M, Fuess H, Vogt T (1991) J Phys Chem 95:5255
71. Czjzek M, Jobic H, Bée M (1991) J Chem Soc Faraday Trans 87:3455
72. Schrimpf G, Tavitian B, Espinat D (1995) J Phys Chem 99:10932
73. Mellot C, Espinat D, Rebours B, Baerlocher Ch, Fischer P (1994) Catal Letters 27:159
74. Mellot C, Espinat D (1994) Bull Soc Chim Fr 131:742
75. Mellot C, Simonot-Grange MH, Pilverdier E, Bellat JP, Espinat D (1995) Langmuir 11:1726
76. Jobic H, Bée M, Méthivier A (1997) Recent research reports, supplementary materials to the 11th Int. Zeolite Conf., Seoul, Korea, Chon H, Uh YS (eds) Hanrimwon Publishing Company, Seoul, p 279
77. Descours A, Méthivier A, Espinat D, Rebours B, Baerlocher Ch, in preparation
78. Czjzek M, Vogt T, Fuess H (1991) Zeolites 11:832
79. Czjzek M, Vogt T, Fuess H (1992) Zeolites 12:237

80. Klein H, Kirschhock C, Fuess HJ (1994) J Phys Chem 98:12345
81. Kirschhock C, Fuess H (1996) Zeolites 17:381
82. Kirschhock C, Fuess H (1997) Microporous Materials 8:19
83. Newsam JM, Silbernagel BG, Garcia AR, Hulme R (1987) J Chem Soc Chem Commun 664
84. Newsam JM (1989) J Phys Chem 93:7689
85. Newsam JM (1987) J Chem Soc Chem Commun (1987) 123
86. Barrer RM, Villiger H (1969) Z Krist 128:352
87. Taylor JC (1987) J Chem Soc Chem Commun 1186
88. Taylor JC (1987) Zeolites 7:311
89. Sacerdote-Peronnet M, Mentzen BF (1993) Mat Res Bull 28:767
90. Mentzen BF (1987) Mat Res Bull 22:337
91. Goyal R, Fitch AN, Jobic H. J Phys Chem, submitted
92. Vigné-Maeder F, Jobic H (1990) Chem Phys Lett 169:31
93. Snurr RQ, Bell AT, Doros NT (1993) J Phys Chem 97:13742
94. Fang MP, Sokol PE, Wang Y (1994) Phys Rev B 50:12291
95. Llewellyn PL, Coulomb JP, Grillet Y, Patarin J, Lauter H, Reichert H, Rouquerol J (1993) Langmuir 9:1846
96. Llewellyn PL, Coulomb JP, Grillet Y, Patarin J, Andre G, Rouquerol J (1993) Langmuir 9:1852
97. Coulomb JP, Martin C, Grillet Y, Tosi-Pellenq N (1994) In: Weitkamp J, Karge HG, Pfeifer H, Hölderich W (eds) Zeolites and related microporous materials: state of the art 1994; Proc. 10th Int. Zeolite Conf., Garmisch-Partenkirchen, Germany, July 17–22, 1994. Elsevier, Amsterdam, Stud Surf Sci Catal 84:445
98. Torrie BH, Brown ID, Petch HE (1964) Can J Phys 42:229
99. Boutin H, Safford GJ, Danner HR (1963) J Chem Phys 39:488
100. Bartl H (1970) N Jb Miner Mh 298
101. Pechar F, Schäfer W, Will G (1983) Z Krist 164:19
102. Artioli G, Smith JV, Kvick Å (1984) Acta Cryst C40:1658
103. Alberti A, Vezzalini G (1981) Acta Cryst B37:781
104. Kirfel A, Orthen M, Will G (1984) Zeolites 4:140
105. Kvick Å, Smith JV (1983) J Chem Phys 79:2356
106. Belitsky IA, Gabuda SP, Joswig W, Fuess H (1986) N Jb Miner Mh 541
107. Smith JV, Pluth JJ, Artioli G, Ross FK (1983) In: Olson D, Bisio A (eds) Proc. 6th Int. Zeolite Conf., Reno, USA, July 10–15, 1983. Butterworth, Guildford, p 842
108. Joswig W, Bartl H, Fuess H (1984) Z Krist 166:219
109. Kvick Å, Ståhl K, Smith JV (1985) Z Krist 171:141
110. Hambley TW, Taylor JC (1984) J Solid State Chem 54:1
111. Pluth JJ, Smith JV, Kvick Å (1985) Zeolites 5:74
112. Alberti A, Vezzalini G, Tazzoli V (1981) Zeolites 1:91
113. Artioli G, Smith JV, Kvick Å (1985) Acta Cryst C41:492
114. Miller SA, Taylor JC (1985) Zeolites 5:7
115. Artioli G, Rinaldi R, Kvick Å, Smith JV (1986) Zeolites 6:361
116. Rinaldi R, Vezzalini G (1985) In: Drzaj B, Hocevar S, Pejovnik S (eds) Zeolites, synthesis, structure, technology and application, Proc. Int. Symp., Portorose, Yugoslavia, Sept. 3–8, 1984. Elsevier, Amsterdam, Stud Surf Sci Catal 24:481
117. Kvick Å, Artioli G, Smith JV (1986) Z Krist 174:265
118. Ståhl K, Kvick Å, Ghose S (1989) Zeolites 9:303
119. Larsson K, Tegenfeldt J, Kvick Å (1989) J Phys Chem Solids 50:107
120. Artioli G, Smith JV, Kvick Å (1989) Zeolites 9:377
121. Kvick Å (1986) Trans Acta Cryst A22:97
122. Jirak Z, Vratislav S, Zajicek J, Bosacek V (1977) J Catal 49:112
123. Jirak Z, Vratislav S, Bosacek V (1980) J Phys Chem Solids 41:1089
124. Bosacek V, Beran S, Jirak Z (1981) J Phys Chem 85:3856
125. Cheetham AK, Eddy MM, Thomas JM (1984) J Chem Soc Chem Commun 1337
126. Czjzek M, Jobic H, Fitch AN, Vogt T (1992) J Phys Chem 96:1535

127. Sun T, Seff K (1992) J Catal 138:405
128. Bull LM, Cheetham AK, Hopkins PD, Powell BM (1993) J Chem Soc Chem Commun 1196
129. Krossner M, Sauer J (1996) J Phys Chem 100:6199
130. Sauer J (1996) Science 271:774
131. Jobic H, Tuel A, Krossner M, Sauer J (1996) J Chem Phys 100:19545
132. Smith L, Cheetham AK, Morris RE, Marchese L, Thomas JM, Wright PA, Chen J (1996) Science 271:799
133. Cockcroft JK (1993) Profil, a multipattern and multiphase Rietveld refinement program with chemical and thermal constraints. Birkbeck College, London
134. Larson AC, Von Dreele RB (1987) GSAS, Los Alamos National Laboratory Report No. LA-UR-86-748
135. Maichle JK, Ihringer J, Prandl W (1988) J Appl Cryst 21:22

Electron Microscopy Studies in Molecular Sieve Science

O. Terasaki

Department of Physics, Graduate School of Science, Centre for Interdisciplinary Research and CREST, Japan Science and Technology Corporation, Tohoku University, Sendai 980-8578, Japan; *e-mail:* terasaki@msp.phys.tohoku.ac.jp

1 **Introduction** 72

1.1 What is a Microscope and What is an Electron Microscope? 72
1.2 Lenses for Electrons 73
1.3 Atomic Scattering Factors 73
1.4 Diffraction and Image of Electrons by Specimens 74
1.5 Scanning and Transmission Electron Microscopes 75
1.5.1 The Scanning Electron Microscope (SEM) 75
1.5.2 The Transmission Electron Microscope (TEM) 75

2 **Why is EM a Good Technique for the Study of Fine Structures of Zeolites?** 77

3 **Electron Microscopy on Zeolites; a Personal Historical View** . . . 78

3.1 Transmission Electron Microscopy 78
3.2 Scanning Electron Microscopy 79

4 **High Resolution Electron Microscopy** 79

4.1 Basic Principle of the HRTEM Image 79
4.2 Resolution 82

5 **Experimental** 84

5.1 Sample Preparation 84
5.2 EM Observation 84

6 **Case Studies of Fine Structures** 84

6.1 High Resolution SEM 85
6.2 High Resolution TEM 87
6.2.1 Fine Structures 87
6.2.1.1 As-Synthesized LTL 87
6.2.1.2 FAU, EMT and Their Intergrowth 88
6.2.1.3 MFI and MEL 89
6.2.2 Surface Structure and Crystal Growth Units 92
6.2.3 Clusters on or in Zeolites 94

Molecular Sieves, Vol. 2

6.2.3.1 Pt/K-LTL 95
6.2.3.2 Metal Compounds in FAU 96
6.2.4 Modulation and Modification of the Framework Structure 97
6.2.4.1 Si/Al Concentration Modulation in MOR 97
6.2.4.2 Dealumination of the Framework 98
6.2.5 New Structure Determination 99
6.2.5.1 BEA 99
6.2.5.2 SSZ-26 and SSZ-33 100
6.2.5.3 ETS-10 100
6.2.5.4 New Porous Materials, MCM and FSM 104

7 **Electron Radiation Damage** 104

8 **Other Topics** 106

8.1 Image Processing 106
8.2 High-Voltage HRTEM 106
8.3 Electron Crystallography 106

9 **Note Added at Revision (24 June 1997)** 107

9.1 Mesoporous Materials 107
9.2 Quantitative Analysis of ED Patterns and HRTEM Images 109

References 110

1 Introduction

1.1 What is a Microscope and What is an Electron Microscope?

The optical microscope is an instrument that uses lenses to produce enlarged images of small objects, especially of objects too small to be seen by the naked eye. It was invented in the late 16th century by a Dutch spectacle-maker. In an optical microscope, the spatial distribution of absorption or reflectance of light is enlarged. Other microscopes are analogous in their functions but employ other radiations/waves than visible light. Enlargement is usually a keyword for a microscope but a wider meaning is to give structural information which is not obtainable by the naked eye such as the X-ray or acoustic microscope. Among different types, an electron microscope (EM) is a microscope in which the output signal is produced by incident electrons, and scanning tunneling microscopes may be included in this category. In this context, two types of EM, i.e. scanning EM (SEM) and transmission EM (TEM), are discussed.

1.2 Lenses for Electrons

Electrons are deflected by magnetic fields by the Lorentz force. Electrons can thus be focused by an electromagnetic lens. It was shown in 1926 that an axial magnetic field acts as a concave lens for incident electrons which are parallel to the axis [1]. This is simply explained by the force in the following. In a magnetic field with rotational symmetry, the radial and axial components (B_r and B_z) of the magnetic field, B, act independently, although there is a relationship, $\partial(rB_r)/\partial r = -r\,\partial B_z/\partial z$, between B_r and B_z because div $\mathbf{B} = 0$. Electrons traveling parallel to the axis will be rotated by B_r. Once a rotational component of electrons is produced they will then be deflected towards the axis by an axial component B_z. It is therefore to be noted that the lens for electrons is always concave not convex and that the aberration cannot be diminished by a combination of the two types of lenses, which is sometimes possible in an optical microscope. In 1931, Knoll and Ruska invented the first EM for which Ruska was awarded the Nobel prize for physics in 1986 [2].

1.3 Atomic Scattering Factors

X-rays are scattered by electrons in an atom. The atomic scattering factor for X-rays, $f_x(s)$ (in Å), is given by

$$f_x(s) = e^2/mc^2 \int \varrho(r)\exp(-is\,r)\,dr = 2.82 \times 10^{-5} \int \varrho(r)\exp(-is\,r)\,dr$$

where s is the scattering vector, and is given by $s = 4\pi\sin\theta/\lambda$ (in Å^{-1}), $\varrho(r)$ is the electron charge distribution and other terms have the standard meaning.

Electrons are scattered by the electrostatic potential formed by the electrons and nuclei of the constituent atoms. The atomic scattering factor for electrons, $f_e(s)$ (in Å), is related to that for X-rays by the Mott formula given by the following equation:

$$\begin{aligned} f_e(s) &= me^2/2h^2\,[Z - f_x(s)]/(\sin\theta/\lambda)^2 \\ &= 2/a_B[Z - f_x(s)]/(4\pi\sin\theta/\lambda)^2 = 0.02393\,[Z - f_x(s)]/(\sin\theta/\lambda)^2 \end{aligned}$$

where Z is an atomic number, and a_B is the Bohr radius (0.529 Å). The wavelength of electrons λ (in Å) is given by de-Broglie at an accelerating voltage E, which is measured in volts:

$$\lambda = (h^2/2me)^{1/2}\,[1/E\,(1 + eE/2mc^2)]^{1/2} = 1.226/[E^{1/2}(1 + 0.9778 \times 10^{-6}\,E)^{1/2}]$$

The differences in diffraction behavior between electrons, X-rays and neutrons are summarized in Table 1.

It is important that the atomic scattering factor for electrons is approximately 10^4 and 10^5 times as large as for X-rays and for neutrons, respectively. This suggests that smaller scatterers can provide sufficient information and thus much smaller objects can be studied as single crystals with electrons. Furthermore, we can make lenses and therefore observe both the electron diffraction (ED) pattern and the image using the EM.

Table 1. Diffraction behavior electrons, X-rays and neutrons

	Principal scatterer	Amplitude (Å) (at s = 0 for Si)	Wavelength (Å)
electrons	electrostatic potential	5.8	0.037 (100 kV)
X-ray	electron [a]	3.9×10^{-4}	1.54 (Cu kα)
neutron	nucleus [b]	4.1×10^{-5}	1.33 [c]

[a] Scattering amplitude for the nucleus is ca. 10^{-3} smaller than for the electron.
[b] Scattering amplitude for the magnetic electron is about the same order.
[c] Room temperature.

1.4 Diffraction and Image of Electrons by Specimens

In electron diffraction, information is accumulated from all the areas irradiated by the incident electrons as shown in Fig. 1. Relative scattering amplitudes for the different reflections are different for each column depending on its structure. Intensity for each reflection is the accumulation from that of different

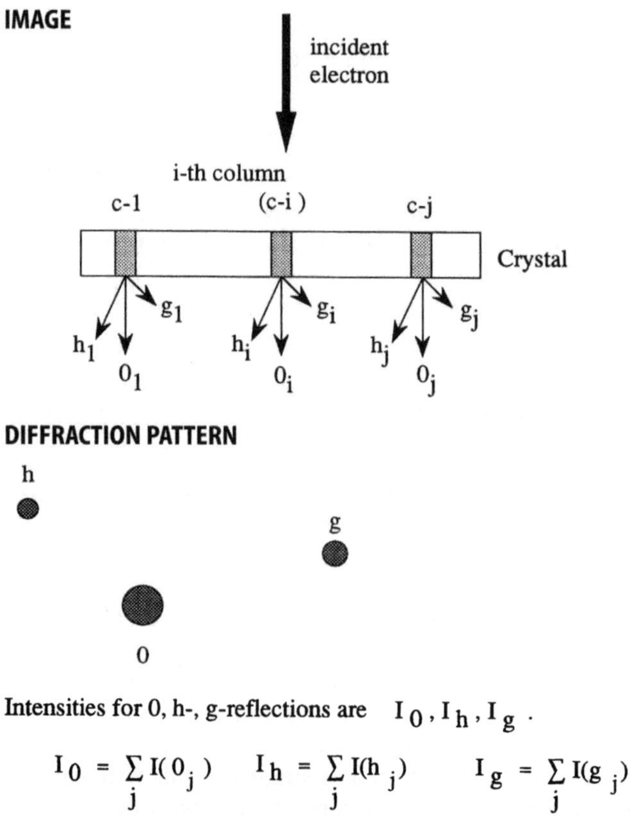

Fig. 1. Schematic diagram of diffraction and image

columns (Σ_j is the summation over j columns). It is therefore necessary to use dark field imaging to identify the place where a particular reflection or diffuse scattering is coming from. However, an EM image itself gives local or spatial information. From the observed image we can obtain the Fourier diffractogram from the specified area in the image by using optical diffraction (Fourier transformation). In this respect, high-resolution transmission electron microscope (HRTEM) images are generally more useful than ED patterns. The big advantage, however, will be obtained by the combination of ED patterns and HRTEM images for electron crystallography (Sects. 8.3 and 9.2). This technique will soon come to the attention of the zeolite community. It is to be noted that the state of clusters in the spaces of zeolites, i. e. size, relative position to the framework and density, is often non-uniform, in this case HRTEM images can give very useful local information.

1.5 Scanning and Transmission Electron Microscopes

1.5.1 *The Scanning Electron Microscope (SEM)*

In the SEM, a source of electrons is focused into a very fine probe that is rastered over the surface of crystals by a scanning coil. A schematic diagram of the SEM is shown in Fig. 2. The output signal can then be used to modulate the brightness of a cathode ray tube (CRT) as a function of the surface position where the electrons are focused. When high energy incident electrons interact with atoms, they will be scattered elastically or inelastically. In an inelastic scattering event, a part of the incident energy is transferred to other electrons in atoms (if the energy transferred exceeds the work function of the crystal, secondary electrons can exit from the crystal) or produces light or X-rays induced by the transition between an excited state and the ground state. Several different kinds of outputs can be used including secondary electrons, back scattered electrons, X-rays, transmitted electrons, beam induced current and light cathodoluminescence. Hereafter, the output of the SEM will be confined to secondary electrons. The secondary electron yields increase with the decreasing glancing angle of the incidence to the specimen surface, i.e. the yields depend on the surface topology. The superiority of the SEM over the optical microscope is the particularly high spatial resolution and the great depth of focus, and therefore we can observe surface-topological information from a large specimen area at high resolution.

1.5.2 *The Transmission Electron Microscope (TEM)*

In the TEM, there are two modes, i. e. information is either in wave vector (reciprocal) space at the back-focal plane or in real space at the image plane. The first corresponds to an ED pattern and the latter to an image. More details will be discussed in Sect. 4.1.

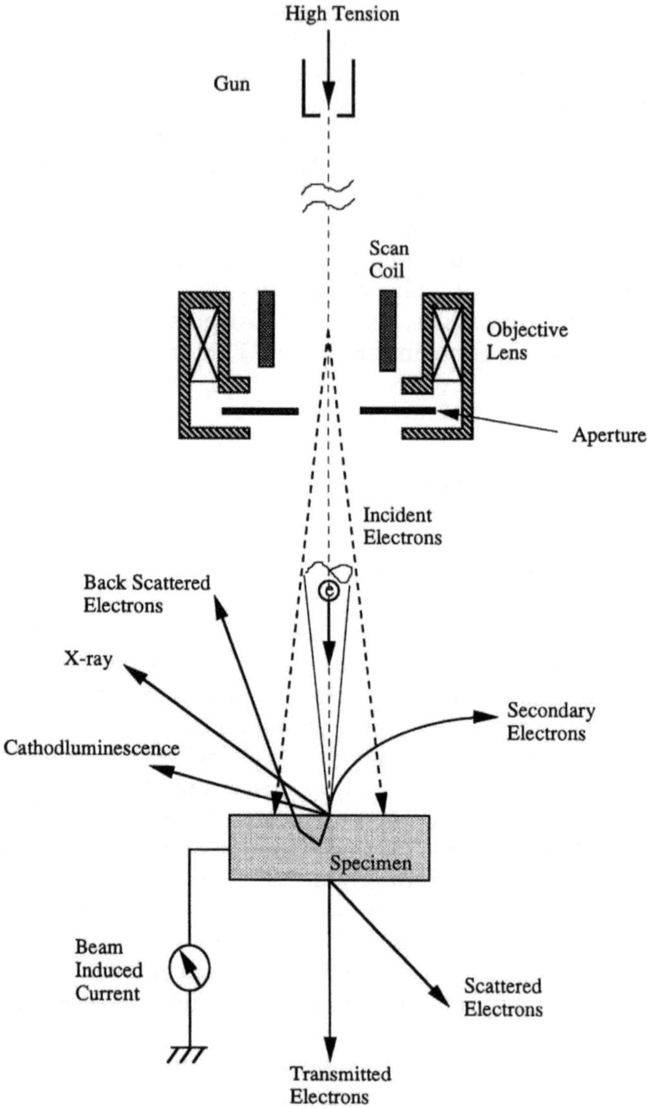

Fig. 2. Schematic diagram of an SEM

There are three independent contrast mechanisms for TEM images, which are: (i) intensity contrast (absorption), (ii) amplitude contrast (diffraction), and (iii) phase contrast (out-of-focus).

Absorption contrast comes mainly from a loss of incident electrons in the transmission beam by multiple inelastic scattering of the electrons, and diffraction contrast comes from the removal of scattered electrons from the image-forming beam, because they fall outside of an objective aperture, hence the objective aperture is frequently called the "contrast aperture". Since zeolite crystals are weak phase objects, and essential structural information is carried out by the reflections within an objective aperture normally used, the latter is the most important for the present subject and will be described in detail in Sect. 4.

2 Why is EM a Good Technique for the Study of Fine Structures of Zeolites?

Zeolites are crystalline aluminosilicate materials with the nominal composition $M_{x/m}[Al_xSi_{1-x}O_2]nH_2O$, where M is a cation of valence m. Their frameworks are built from corner-shared TO_4-tetrahedra (T stands for Si or Al) to produce channels or cavities (hereafter called spaces) of molecular dimensions inside the crystals [3]. Zeolites are mostly synthesized as particles of ca. 1 μm in size. Briefly summarized below are the five major advantages for using EM for the study of zeolites, which will be discussed in more detail in Sect. 6:

(i) If zeolites are free from defects, their structures can be determined by X-ray powder diffraction. However, they may often contain different types of defects and, moreover, their chemical and physical properties can heavily deviate from their expected ideal structures. It is therefore essential to characterize any defects which may occur, as reported for the case of LTL [4–7].

(ii) Zeolites form many different families in which common structural building units are found in the frameworks. They have a strong tendency to form intergrowths between different structures within the same families, such as ERI/OFF in ABC-6 [8,9] and EMT/FAU in the FAU family [10–12]. There are also many polymorphs where structures are well described by the stacking sequence of the sheets such as BEA [13] and the titanosilicate ETS-10 [14–16]. To characterize these, electron microscopy is the most useful technique.

(iii) In order to synthesize high quality or novel zeolites, two factors are very important to know: (a) what kind of molecular units are the elemental unit of the structure for growth, and (b) how to control the connection of the unit by using structure directing agents. If new methods to control and supply the unit, for example by CVD (chemical vapor deposition), could be found then it might be possible to grow larger single crystals. Therefore a knowledge of the surface structures of zeolites is very important both for understanding crystal growth processes expected in order to synthesize high quality or novel zeolites and for understanding catalytic reactions which may occur on the surfaces more quantitatively. Examples of this type are found in FAU [6, 12, 17] and LTL [18] systems.

(iv) Zeolites have been used as containers to make new "quantum materials" in their spaces [19–23], and also used as complex catalysts by the incorporation or suspension of clusters of metals or compounds. In these cases, it is vitally important to determine the size of the clusters and locate their positions, whether they are in the spaces of the zeolites or on the external surface, and also to check whether the crystallinity of the zeolites is retained after the treatment [6, 7, 24, 25]. Electron microscopy provides the crucial starting point for characterizing the fine structures and solving such complex problems. Fine details of the structure and the nature of disorder can be directly revealed by HRTEM images.

(v) A crystal with high crystallinity shows mostly a regular crystal morphology which is either an equilibrium form or a growth form, even though the crystal size is small, (ca. 1 μm). Both forms must be commensurate with the point symmetry to which the crystal belongs. It is very informative that natural zeolites show beautiful crystal shapes and have very good crystallinity. Thus we can obtain some information about the *crystallinity* from the external morphology. Zeolites are normally synthesized under conditions of high supersaturation, and therefore a growth form rather than an equilibrium form is observed. The growth form is governed by the anisotropy of the growth rate which is sensitive to the growth conditions. From experience we have come to the conclusion that there is a strong correlation between the thermal stability and the stability of crystals under an electron beam, and that electron stability of zeolites is strongly dependent on the crystallinity. Therefore, a few years ago at the Annual Meeting of the Japanese Zeolite Association we proposed that the simplest way to find synthesis conditions for a crystal (in the same framework structure and the same Si/Al ratio) with a high thermal stability was to observe a crystal with a nice morphology by SEM irrespective of crystal size, and this is a very apparent indication of high crystallinity.

There are two major disadvantages for the study of zeolites by TEM:

(i) Zeolites are quite electron beam sensitive, (this is the most serious [26])
(ii) HRTEM images suffer from artificial electron optical effects [27–29]; these will be discussed briefly later.

In this review, I will describe the history of EM studies on zeolites in Sect. 3, the basic principles in Sect. 4 and show some examples on the fine structures of zeolites in Sect. 6. On the whole I will try to show in what way electron microscopy is a powerful and invaluable technique for the study of the fine structures of zeolites and related materials.

3
Electron Microscopy on Zeolites; a Personal Historical View

3.1
Transmission Electron Microscopy

Menter first observed lattice fringes of platinum phthalocyanine by EM in 1956 [30] and then observed two sets of lattice fringes of {111} type of natural FAU with ⟨110⟩ incidence [31]. These two specimens have large lattice spacings which should be larger than the resolution limit (ca. 10 Å at that time). However, it is astonishing that he recorded the lattice fringes from those specimens, since they are now known to be quite electron beam sensitive materials. These images were basically two- or three-beam interference fringes (lattice fringes). After Menter's pioneering work, EM was mostly applied to studies of zeolite fine structure by obtaining ED patterns from small crystals for LTL, OFF and ERI [32]. The authors stated that "As X-ray powder diffraction is not easily able to detect the presence of small amounts of impurities, the samples were examined in the electron microscope" [32] and that "The importance of electron diffraction was also shown" [33]. As far as I know, imaging was first applied independently to the study by two groups. One was the direct observation of planar faults in ERI [34], and the other was the observation of a lattice image of LTL in 1972 [35].

In the early seventies, Sanders and co-worker developed the image from a lattice image to the structure image [36], with which I am concerned in this article. Later he applied the technique for the study of the fine structure of zeolites with a deep understanding of electron microscopy [37, 38]. In late 1970s, Thomas and his group in Cambridge developed many techniques useful for HRTEM observation of zeolites, such as to modify zeolites in order to improve their life under the electron beam by removing zeolitic water or replacing Al by Si (dealumination), to give higher contrast by ion exchange with heavy ions and to use computer modeling and simulation, and elaborated the techniques presently available for the study [26, 39–41].

3.2 Scanning Electron Microscopy

Most zeolite studies with SEM have been carried out with natural zeolites. One publication [42] gives encyclopedic information on the natural zeolites but, compared to recent technical improvements, no basic studies have been found.

4 High Resolution Electron Microscopy

4.1 Basic Principle of the HRTEM Image

The velocity of incident electrons will be changed in a crystal due to the electrostatic potential and they must satisfy continuous boundary conditions at both the entrance and exit surfaces of the crystal. We will take the direction of incident electrons to be the z-axis. The role of the objective lens is to transfer the Fraunhofer diffraction pattern produced by the crystal at infinity to the back focal plane of the lens, i. e. to give the Fourier transform of the electron wave field at the exit surface of the crystal [transmission function $\phi(x, y)$] at the back focal plane. Electron diffraction patterns are observed as an intensity distribution in reciprocal space. The lens further Fourier transforms the diffraction pattern to an image plane (EM images are observed as an intensity distribution in real space) as schematically shown in Fig. 3. As a first approximation, the crystal is treated as a periodic object which will modify only the phase of the electron wave function, (weak phase object approximation).

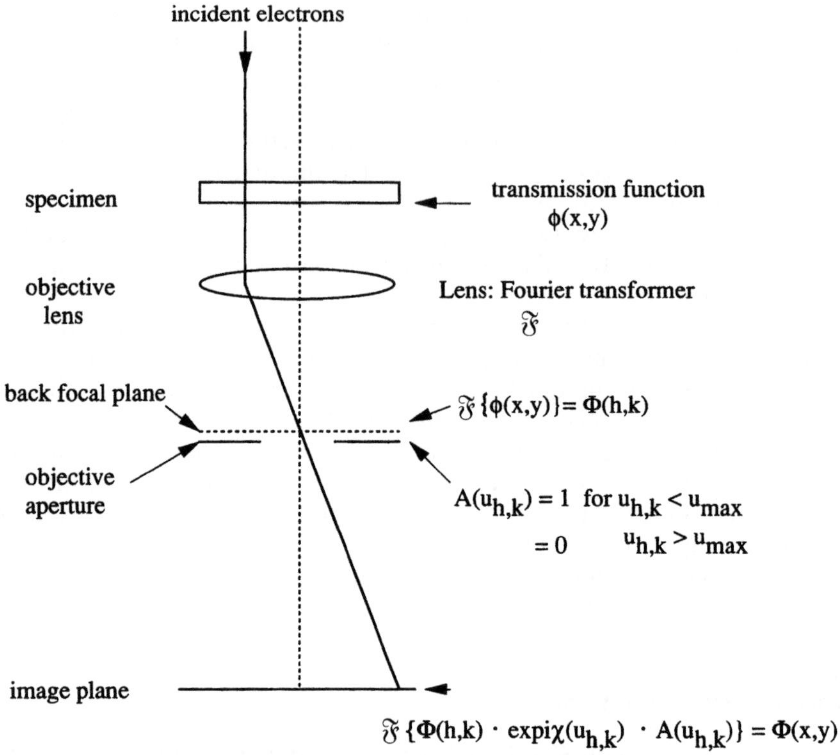

Fig. 3. Schematic diagram of HREM imaging

From analogy with light optics, the refractive index of the crystal for electrons, n, is given by:

$$n = \lambda_{vac}/\lambda_{cryst} = \mathrm{SQRT}\,[(E + \phi)/E] \approx 1+\phi/(2E)$$

where λ_{vac} and λ_{cryst} are the electron wavelengths in a vacuum and in the crystal, respectively, E is the accelerating voltage and ϕ is the mean crystal potential. The phase shift, θ, is the phase difference between the electrons traveling in the crystal and a vacuum of thickness t and is given by

$$\theta = 2\pi(n-1)\,t/\lambda \approx \pi\int_0^t \phi(x,y,z)\,dz/\lambda E \equiv \sigma\,\phi_p(x,y)$$

where σ is the interaction parameter of the electrons with the crystal, $\sigma = \pi/\lambda E$. The projected potential, $\phi_p(x,y)$, is the crystal potential of the crystal with thickness t and $\phi_p(x,y) = \int_0^t \phi(x,y,z)\,dz$ (in volt · Å). For simplicity we assume that the crystal has a center of symmetry and $\phi_p(x,y)$ is taken to be real.

The electron wave field at the exit surface, i.e. the transmission function $\phi(x, y)$, will be:

$$\phi(x,y) = \exp[-i\sigma\phi_p(x,y)]$$

where $|\phi(x,y)|^2 = 1$.

From the assumptions of the weak phase object, that is $\sigma\phi_p(x,y) \ll 1$, then $\phi(x,y) \approx 1 - i\sigma\phi_p(x,y)$. If the lens is perfect and there is no defocus, the Fourier transform, $\mathfrak{F}$, of the transmission function, $\phi(x,y)$

$$\mathfrak{F}\{\phi(x,y)\} \equiv \Phi(h,k) = \delta(u_{0,0}) - i\sigma\mathfrak{F}\{\phi_p(x,y)\} = \Phi(u_{0,0}) + \Phi(u_{h,k})$$

will be formed at the back focal plane, where $u_{h,k}$ is the wave vector for a reflection (h,k) and is given by the equation $u^2_{h,k} = (h/a)^2 + (k/b)^2$ for an orthorhombic crystal. The first term corresponds to the transmission wave (0,0) and the second to the reflections (h,k). It should be noted that the phase of the diffracted beams is changed by $\pi/2$ from the above equation. $\mathfrak{F}\{\phi(x,y)\}$ is proportional to and derived from the crystal structure factor for electrons, $F_e(h,k)$. The wave field at the image plane, $\Phi(x,y)$, is given by the $\mathfrak{F}$ of $\Phi(h,k)$, and the image (intensity) is given by $I(x,y) = \Phi(x,y)\Phi^*(x,y)$ which is uniform, i.e. there is no contrast.

The objective aperture will take the beams, passing through the aperture, for image formation. We introduce an aperture function, $A(u_{h,k})$, where $A(u_{h,k}) = 1$ for $u_{h,k} < u_{max}$ and $A(u_{h,k}) = 0$ for $u_{h,k} > u_{max}$.

Fortunately, the objective lens introduces an additional phase shift depending on the direction of the diffracted beams, $u_{h,k}$, the amount of defocus f and the spherical aberration C_s of the lens. When the incident electrons are parallel to the EM optical axis, the phase of the reflection (h, k) will be changed by the lens relative to the direct beam by $\chi(u_{h,k})$, where:

$$\chi(u_{h,k}) = 2\pi(\Delta f \cdot \lambda \cdot u^2_{h,k}/2 + C_s \cdot \lambda^3 \cdot u^4_{h,k}/4)$$

$\exp\{i\,\chi(u_{h,k})\}$ is known as the contrast transfer function, CTF, which will be discussed later.

Therefore the wave field, $\Phi(x,y)$, at the image plane is given by:

$$\begin{aligned}\Phi(x,y) &= \mathfrak{F}\{\Phi(h,k) \cdot \exp\{i\,\chi(u_{h,k})\} \cdot A(u_{h,k})\}\\ &= 1 - i\sigma \cdot \phi_p(-x,-y) * \mathfrak{F}\{A(u_{h,k}) \cdot \exp\{i\,\chi(u_{h,k})\}\}\end{aligned}$$

where $*$ is the convolution operation.

To give contrast, the imaginary part of $\mathfrak{F}\{A(u_{h,k})\} \cdot \exp\{i\chi(u_{h,k})\}$ and therefore $\sin\chi(u_{h,k})$ plays an important role, since $\exp\{i\chi(u_{h,k})\}$ is $\cos\chi(u_{h,k}) + i\sin\chi(u_{h,k})$. Sin $\chi(u_{h,k})$ is shown in Fig. 4 for a JEOL 4000 EX EM.

The intensity distribution obtained as an HRTEM image will therefore be:

$$I(x,y) = \Phi(x,y)\,\Phi^*(x,y) \approx 1 + 2\sigma \cdot \phi_p(-x,-y) * \mathfrak{F}\{A(u_{h,k}) \cdot \sin\chi(u_{h,k})\}.$$

The function $\mathfrak{F}\{A(u_{h,k}) \cdot \sin\chi(u_{h,k})\}$ is a function with a negative value that has a sharp peak. The CTF plays a similar phase-changing role in the optical

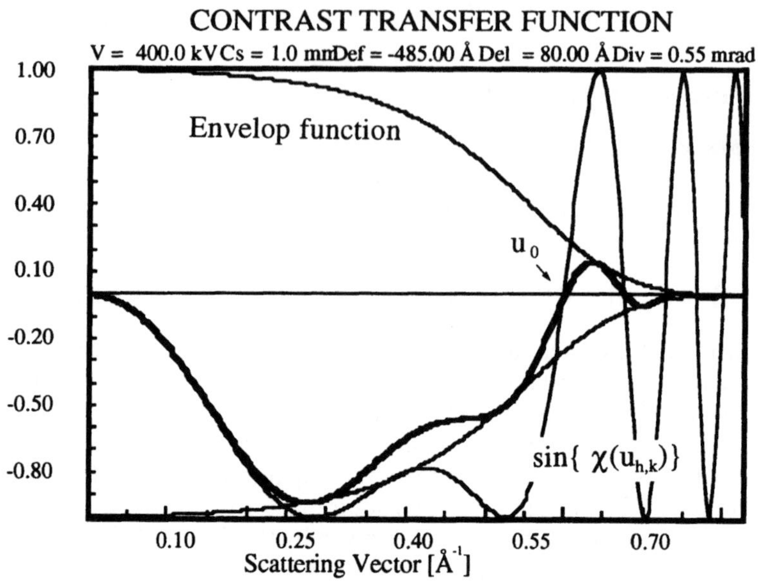

Fig. 4. Imaginary part of CTF, sin $\chi(u_{h,k})$ for a 4000 EX at optimum focus condition (Δf = -485 Å). The *thin curves* are sin $\chi(u_{h,k})$ and envelop a function due to chromatic aberration (not discussed); the *thick curve* is the product of their multiplication

microscope to Zernike's phase plate of $\lambda/4$ [43]. If the crystal has no center of symmetry and $\phi_p(-x,-y)$ has an imaginary part, $\phi^i_p(-x,-y)$, then part of cos $\chi(u_{h,k})$ will give an effect as a correction.

We have come to the very important conclusion that the contrast of HRTEM images is proportional to the projected potential of the crystal, which is smeared by convolution with $\mathfrak{F}\{A(u_{h,k}) \cdot \sin\chi(u_{h,k})\}$, the so-called impulse response function. This is the key to obtaining structural information from HRTEM images.

4.2 Resolution

The resolution of any optical microscope is limited by the diffraction aberration, δ_d, as $\delta_d = 0.61\lambda/\alpha$, where α is an effective aperture angle known as the Rayleigh criterion. It is possible to reach $\alpha \approx 1$ for a light optical microscope, therefore the attainable resolution is around the wavelength of light. According to Table 1, the wavelengths of electrons are far less than 1 Å. Can we obtain such a high resolution (to the wavelength of an electron) in an EM? What are the major factors for limiting the resolution of the TEM and SEM?

(i) For the TEM,in contrast to an optical microscope, the resolution of the TEM is limited not by the wavelength but by the spherical (and chromatic) aberration of the objective lens through the contrast transfer function CTF. This is because the resolution is given not only by the Rayleigh criterion but also by the effect of the aberration (δ_{ab}) of the lens, as:

$$\delta = \mathrm{SQRT}(\delta_d^2 + \delta_{ab}^2) \approx \delta_{ab}[1 + \delta_d/(2\,\delta_{ab})]$$

since $\delta_{ab} \gg \delta_d$ as is seen in the following. The CTF was given in the previous section as: CTF = exp $\{i\,\chi(u_{h,k})\}$, where $\chi(u_{h,k}) = 2\pi(\Delta f \cdot \lambda \cdot u_{h,k}^2/2 + C_s \cdot \lambda^3 \cdot u_{h,k}^4/4)$.

An optimal focus is given by the condition when the first zero cross and $\sin\{\chi(u_{h,k})\}$ values have maximum values in the widest region for $u_{h,k}$, and is known as the Scherzer focus, defined as $\Delta f = -(C_s\lambda)^{1/2}$.

The resolution (δ_{ab}) obtainable in this condition is $0.43\,(C_s\lambda^3)^{1/4}$ from the inverse value of u_0 at the first zero cross of CTF at the focus condition (see Fig. 4).

In order to obtain better resolution in HRTEM, one should (i) reduce the wavelength, (ii) reduce the spherical aberration coefficient, and (iii) image restoration by correcting the functional dependence of CTF on $u_{h,k}$. The first two factors are obvious from the functional dependence of CTF on them. Roughly speaking, C_s reduces as the focal length of the objective lens decreases, that is, as the excitation of the objective lens increases, and the values of C_s and the focal length become almost equal. It is more effective to reduce λ for better resolution as is given by $0.43\,(C_s\lambda^3)^{1/4}$. Information from the area beyond the zero crossing is also useful if we take the phase change into account. Using a highly coherent electron beam, there have been attempts to extend the resolution higher by recovering information from this oscillating area (focal series and tilt series reconstructions).

(ii) The resolution of an SEM is essentially limited by the electron probe size on the specimen. Electrons from a source of size s_0 are de-magnified through condenser lenses by some factor M to a probe at the specimen surface of size d_0. Other independent factors also blur the size, and the effective probe size, d, is given by:

$$d^2 = d_0^2 + d_s^2 + d_c^2 + d_d^2$$

where $d_s \sim C_s\alpha^3$ is the size produced by spherical aberration (C_s) of an objective lens, $d_c \sim C_c\alpha\,(\Delta E/E)$ due to chromatic aberration (C_c) of the lens (E and ΔE are incident energy and energy spread) and $d_d \sim C_c(\lambda/\alpha)$ is a spread due to diffraction of an aperture, respectively. $\Delta E/E$ becomes larger for a lower accelerating voltage of incident electrons. This is further blurred (convoluted) by the size of the specimen volume giving rise to the signal, i.e. lengths of diffusion of incidence and of escape of secondary electrons. By developing a very bright coherent electron source together with smaller aberrations of the objective lens, we can make the electron probe size smaller and smaller and the resolution better and better. Now it is possible to reach a resolution in the SEM image using secondary electrons from 10 to 15 Å with the use of field emission guns (FEG).

5
Experimental

5.1
Sample Preparation

Zeolite crystals were dispersed in acetone either by crushing in an agate mortar or by use of ultrasound and subsequently placed on a micro-grid for TEM observation. Removing zeolitic water from the crystals helps to lengthen their lives under the electron beam.

For SEM observations, the powders were directly dispersed on a conductive tape or dispersed in acetone using ultrasound and drops were placed on the tape. With this treatment, as no crushing is employed, the morphology of the crystals will not be altered from the original form.

5.2
EM Observation

Since zeolites have large unit cells, basic structural units and large void spaces, most of the structure can be determined from the arrangement of the spaces. The resolution necessary for the study of fine structures of zeolites is normally not so high as required for other systems and a resolution of around 2.5 Å is sufficient. It is essential to reduce the number of electrons and/or to reduce electron density as much as possible during observation. The time for tilting the crystal to the proper orientation is much longer than that for image recording, and special care must be taken to ensure that ED patterns can be observed with much smaller density. It is very important for the EM to be equipped with a TV system for adjusting electron optical systems, especially correcting astigmatism and finding the best focus conditions for the objective lens just before HRTEM image recording. For recording media, there are several possibilities, (i) ordinary film, (ii) CCD camera (slow-scan), and (iii) imaging plates. For ordinary film, it is better to use higher sensitivity film, although making the sensitivity of film higher and the grain size of photographic emulsion smaller are contradictory, the grain size is still enough to resolve the fine structure of a zeolite even at low magnification.

For HRTEM recording, it is better to choose a lower magnification, at which information about the structure concerned is attainable, normally between 100,000 and 150,000 in order to reduce the electron density required for recording.

6
Case Studies of Fine Structures

Most of the HRTEM and ED patterns were taken with (i) a 400 kV EM (JEM 4000) which has a spherical aberration coefficient (C_s) of 1.0 mm and (ii) a 1 MV EM with C_s = 11.0 mm. ED patterns will be discussed under each problem.

6.1
High Resolution SEM

Scanning EMs equipped with field emission guns (FEG) have provided an enormous advantage especially in obtaining high-resolution images from non-conductive materials such as zeolites. Using a low accelerating voltage has two advantages; firstly in producing selective information from the surface region because of the small penetration depth for low energy electrons, and secondly the ratio of incident to secondary electrons is close to 1 and therefore "charging problems common for insulators" are not serious as the accelerating voltage decreases. However, in order to obtain high-resolution images at low accelerating voltage, an electron source with high brightness and coherence is essential (FEG satisfies these requirements). It is also important to reduce the chromatic aberration of the objective lens especially for a low accelerating voltage.

Here, we confine ourselves to the case of high resolution SEM (HRSEM) images. Several examples are shown in Fig. 5. These are low magnification images of

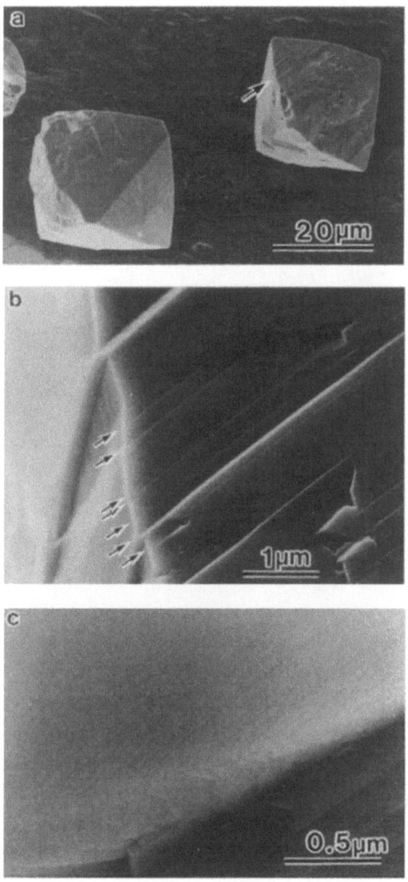

Fig. 5 a–c. Scanning EM images of FAU

a single crystal of FAU (Fig. 5a), and a high magnification image (Fig. 5b) of a part of the right-hand crystal indicated by an arrow in Fig. 5a. In Fig. 5b, arrows show twins in FAU. Figure 5c shows surface steps observed on one of the {111} surfaces of a single crystal of FAU; despite the low contrast, the steps are discernible. Low and high magnification images of -CLO are shown in Figs. 5d, e. In the images craters of rectangular parallelepiped shape on the {100} or trigonal pyramids on {111} can be seen, as marked by large and small arrows, respectively. The parallelepiped-shaped voids are also observed in HRTEM images in which no discernible distortion of the framework structure was observed [44]. Figure 5f shows an SEM image of LTA, on which FAU crystals have overgrown.

Fig. 5d–f. Scanning EM images of -CLO (**d, e**) and LTA (**f**)

6.2
High Resolution TEM

6.2.1
Fine Structures

6.2.1.1
As-Synthesized LTL

LTL has a unit cell of a = 18.4 Å, c = 7.5 Å (space group P6/mmm), and contains a one-dimensional channel with 12-membered rings ca. 7 Å in diameter. The channels are well separated by the framework atoms and therefore very attractive for making one-dimensional materials in the channels. Figure 6 shows a schematic drawing of a projection of the framework structure along [001] (a) and an HRTEM image along [001] (b). All channels shown in Fig. 6a are observed distinctly in the image. A single planar defect, which is not parallel to the channel, may destroy the one-dimensional character. Indeed, it is quite common in LTL to find the growth of one crystal onto another [4–7]. Although the orientation relationship between the crystallites is easily determined by electron diffraction (ED) patterns, HRTEM observation is essential to determine the spatial relationship between them, whether they have blocked each other's

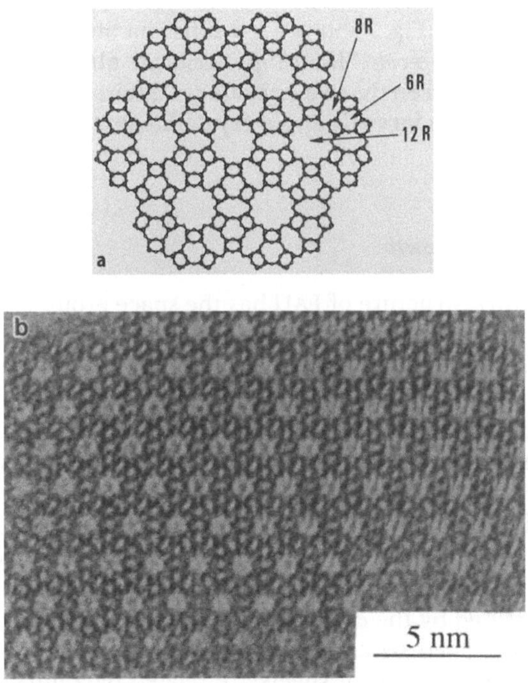

Fig. 6a, b. A schematic drawing of the framework structure of LTL **a** viewed along [001] and **b** a corresponding HRTEM image of LTL taken on a 400 kV EM

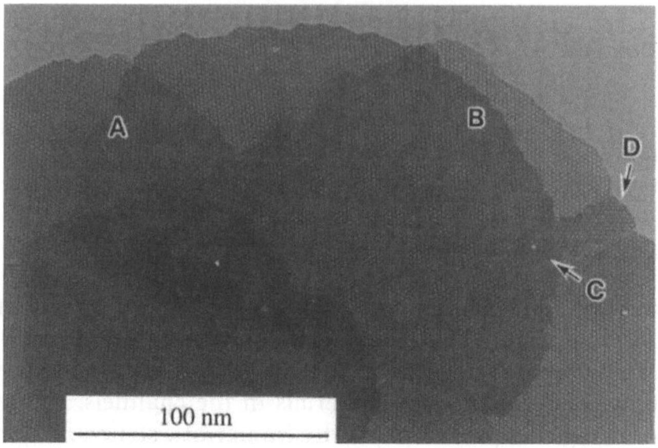

Fig. 7. HRTEM images of LTL taken on a 400 kV EM with [001]. Moiré effects are due to overlapping crystals with rotation angles of ca. 10° (*A*) and 4° (*B*). A coincidence boundary of R32.2° (*C*) and a tilt boundary of ca. 30° (*D*) are also observed

channels or not. We have reported coincidence boundaries of $\sqrt{13} \times \sqrt{13}$ R 32.2° where one crystal grows on another with a rotation of ca. 32° (a non-multiple of 60°) along the c axis, i.e. the channel direction [4]. More than 90% of the one-dimensional channels are blocked in the overlapped region.

The HRTEM image in Fig. 7 shows a few different boundaries or faults in LTL in the same film. Two different "Moiré" patterns are observed in the image, that is, rotational angles between the two crystals are different at the regions of A and B. The coincidence boundary (C) and a crystallite grown in a wedge shape (D) can also be observed.

6.2.1.2
FAU, EMT and Their Intergrowth

The ideal framework structure of FAU has the space group $Fd\bar{3}m$ with a lattice parameter a = 24.7 Å (ignoring the difference between Si and Al). The structure is described by sodalite cages which are linked via D6R to four other sodalite cages in such a way that all the sodalite cages are related by inversion at the centers of the D6Rs, while the framework of EMT has hexagonal symmetry with the lattice parameters a = 17.4 Å and c = 28.4 Å. This was first synthesized by Delprato et al. using the crown ether 18-crown-6 as a structure directing template [45]. Similar to the FAU framework, EMT is constructed from sodalite cages connected through D6Rs; however, one of the four connections, viz. along the c-axis, is related by a mirror operation. Therefore structures of both FAU and EMT can be described by the common structure unit the "faujasite sheet". If a pair of sheets are related by an inversion center then the FAU framework is obtained. Conversely, if a pair of sheets are related by a mirror plane then the EMT framework is obtained. Schematic drawings of the sheet and framework structures of FAU along $[110]_c$ and EMT along $[100]_h$ are shown in Fig. 8 (sub-

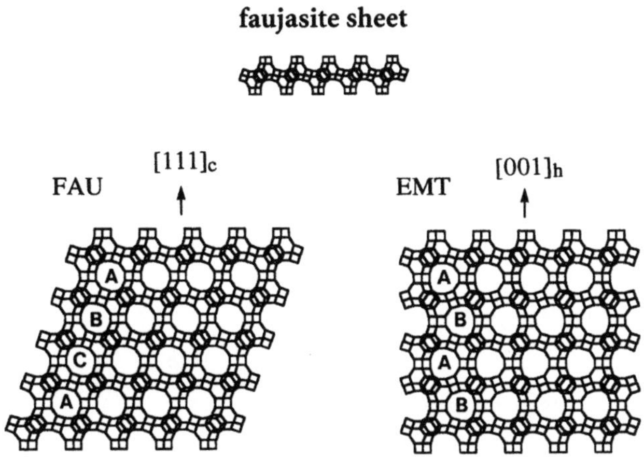

Fig. 8. Schematic drawings of the "faujasite sheet" (*above*). Stacking of channels for FAU; *left* viewed along $[110]_c$; *right* EMT along $[100]_h$

script c and h refer to the cubic and hexagonal structure, respectively). In intergrowth structures, these two directions are coincident. Since channels are imaged as white dots in HRTEM images at the optimum focus condition (see Sect. 4), it is clear from this drawing that the stacking of white dots shows one to one correspondence to that of the sheets. The stacking of white dots is ABCABC for FAU along the $[111]_c$ direction, while that of EMT is ABAB along the $[001]_h$ (see footnote in [11,12] for a more precise structural description).

In FAU there are four independent $\langle 111 \rangle_c$ directions along which EMT may intergrow, but only one direction out of the four is chosen in the synthesis using the mixture of crown-ethers of 18-crown-6 and 15-crown-5 [10–12]. High resolution TEM images of FAU, EMT and their intergrowth are shown in Fig. 9. During this study, we observed oscillatory growth of FAU and EMT, which may be related to the Liesegang ring of gel formation. Figure 10 shows an example of oscillatory growth of the intergrowth in an extreme case. There are many crystallites, and all the EMT are sandwiched by thin FAU layers. An enlarged image of a part of Fig. 10a (indicated by an arrow) is shown in Fig. 10b.

6.2.1.3
MFI and MEL

Two types of crystal, ZSM-5 and ZSM-11, which belong to MFI and MEL types, respectively, were reported at almost the same time [46, 47]. However, the direct evidence of the existence of pure MEL-type has been obtained only quite recently [48], although there were very interesting reports on post synthetically dealuminated high silica ZSM-11 by Fyfe and his colleagues [49]. The characteristics of MFI and MEL-type structures are summarized below:

MFI Type. Figure 11a–c shows projections of the framework structure of MFI along [100], [010] and [001], respectively. One can clearly observe the two-fold rotational axes at the points marked by dots in the projection along [010] in

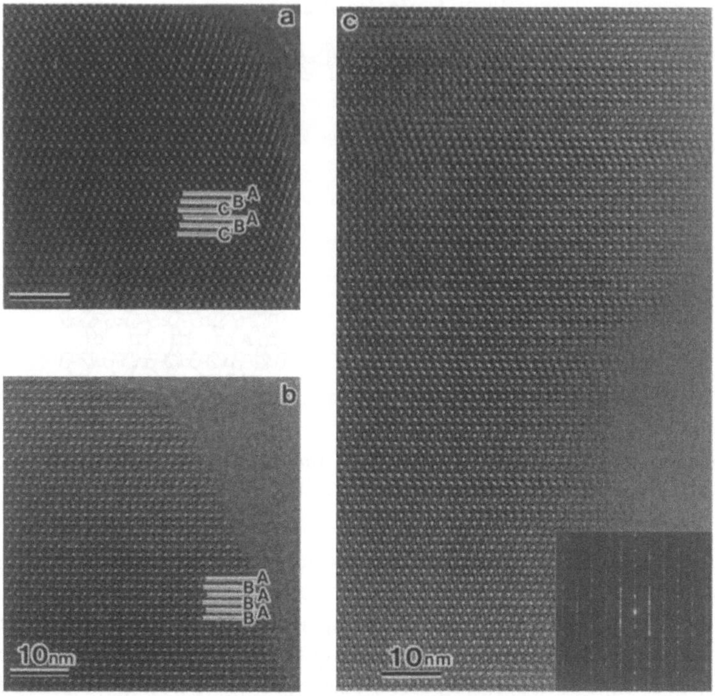

Fig. 9a–c. HRTEM images **a** of FAU taken with$[110]_c$ and **b** EMT with $[001]_h$ and **c** their intergrowth

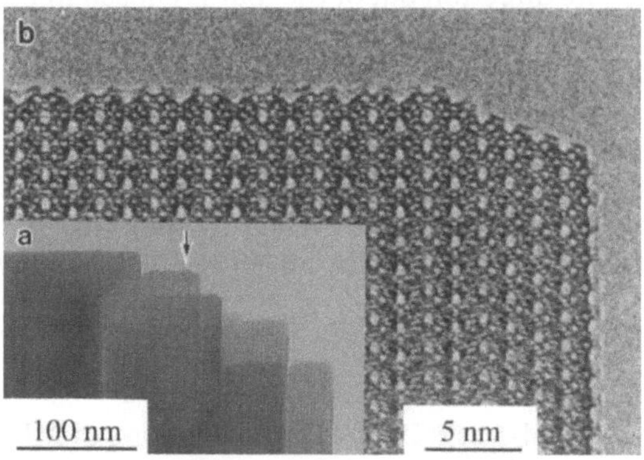

Fig. 10a, b. An HRTEM image of EMT/FAU taken on a 400 kV EM with $[110]_c$ or $[100]_h$

Fig. 11b, which correspond to inversion centers for three dimensions, and also the mirror plane parallel to (010) in the [001] projection (Fig. 11c). Figure 11d–f shows simulated electron diffraction (ED) patterns (for a specimen thickness of 100 Å) corresponding to Fig. 11a–c, respectively, by taking dynamic scattering of electrons into account.

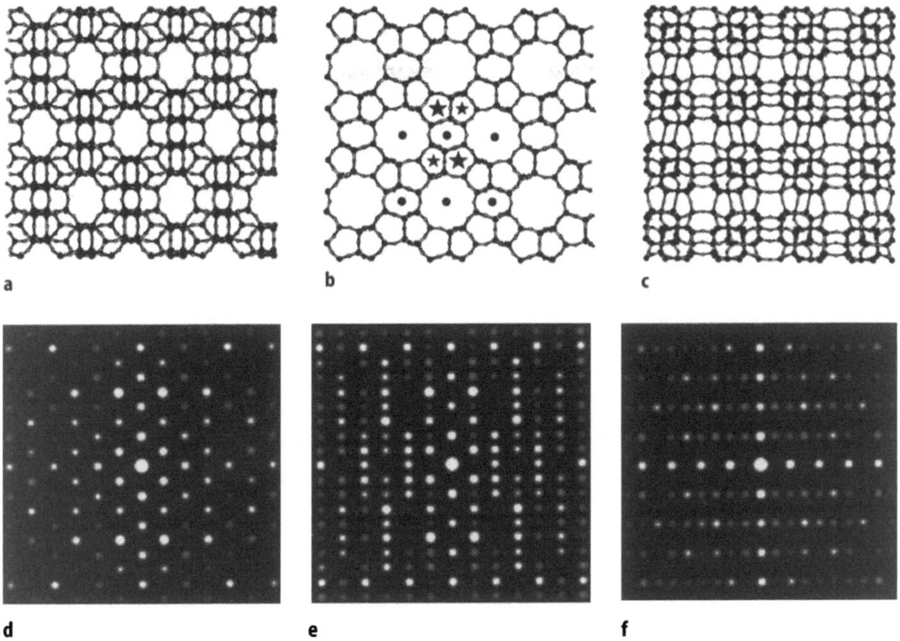

Fig. 11 a–f. Projections of the framework structure of MFI-type along [100] (a), [010] (b), and [001] (c). Simulated ED patterns (d–f) for a, b and c, respectively

MEL Type. Figure 12a, b shows the framework structure of MEL along [100] and [001]. Since MEL belongs to the tetragonal system, the [100] and [010] directions are equivalent, and there is a four-fold axis along the [001] direction. Figure 12c, d shows simulated ED patterns along the [100] and [001] directions, respectively.

How To Differentiate the Two Types. Comparing Figs. 11 and 12, it is very clear that the [001] incidence is the best orientation to observe the difference between the two types. Since the symmetry of the projection along the [001] is p2gm for MFI and p4mm for MEL, either the ED pattern or the HRTEM image alone is sufficient to differentiate between the two. An HRTEM image of MEL taken with [001] is shown together with the ED pattern in Fig. 13. The arrows in the ED pattern show 200 and 020 reflections, both are equivalent by symmetry. Both the HRTEM image and the ED pattern show clearly, by exhibiting 4-fold symmetry, that pure MEL-type has been synthesized. In the other two principal directions [100] and [010], we must observe both an ED pattern and an HREM image as a set in order to demonstrate that the crystal is the MEL-type. The ED pattern in Fig. 11d is very similar to that in Fig. 12c, but the corresponding projected framework structures are quite different as shown in Figs. 11a and 12a. In order to observe the small difference between Fig. 11b and Fig. 12a by HREM images, it is necessary to distinguish between the two different types of 5-rings, which are indicated by two different stars in the figures, due to their different contrast

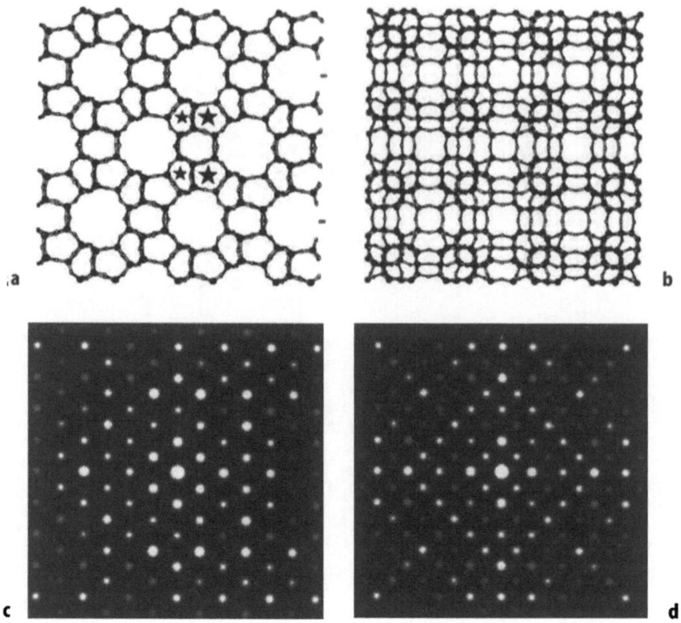

Fig. 12a–d. Projections of the framework structure of MEL-type along ⟨100⟩ ([100] or [010]) (**a**) and [001] (**b**). Simulated ED patterns(**c** and **d**) for **a** and **b**, respectively

or size [50, 51]. Unfortunately, this is not an easy task, and, therefore, the conditions under which to observe the difference in symmetry by an ordinary EM are discussed elsewhere [51].

6.2.2
Surface Structure and Crystal Growth Units

The surface structure is an important factor to be studied for a better understanding of the chemical reaction occurring at the surface. Observation of the growth steps on an atomic scale will provide vital information about the crystal growth mechanism. Figure 14 shows HRTEM images to illustrate growing steps at the surface, i.e. the surface structure of FAU taken along [110]. The height of these steps on the crystallite surface is basically equal to the spacing of $(111)_c$ FAU, i.e. one faujasite sheet shown in Fig. 8. The image is sufficiently clear to allow determination of the nature of the terminating structures at the crystal faces by comparing them with simulated images. The following three different terminating surface models for FAU were simulated by use of the multislice method: (I) with an incomplete sodalite cage, (II) with a complete sodalite cage, and (III) with a complete D6R. These models are shown schematically in Fig. 15. The top is the crystal surface. The simulated images of the type I surface fit very closely to the images of Fig. 14 and they are inserted in the figures, while another HRTEM image of FAU, such as Fig. 10, synthesized by a different

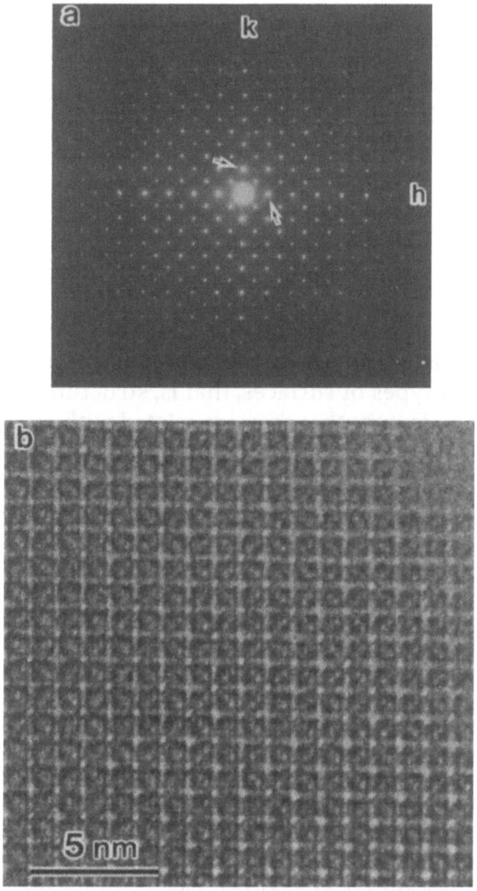

Fig. 13 a, b. An ED pattern (**a**) and an HRTEM image (**b**) of MEL taken with [001]

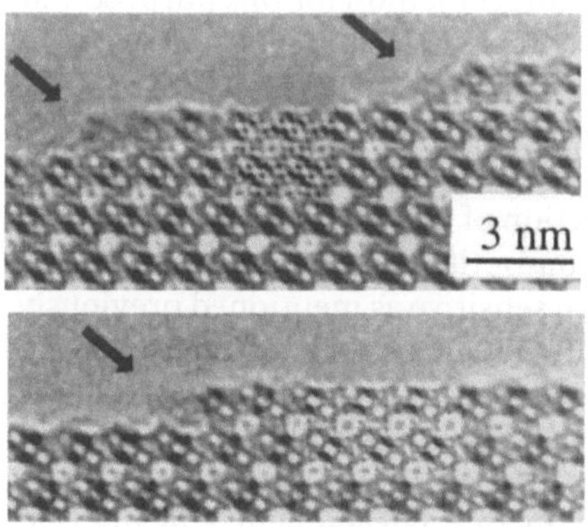

Fig. 14. High resolution TEM surface profile images of $\{111\}_c$ of FAU taken on a 400 kV EM along $[110]_c$. Simulated images (specimen thickness of 50 Å) for the surface model I and focus of –900 Å (*above*) and –700 Å (*below*) are inserted

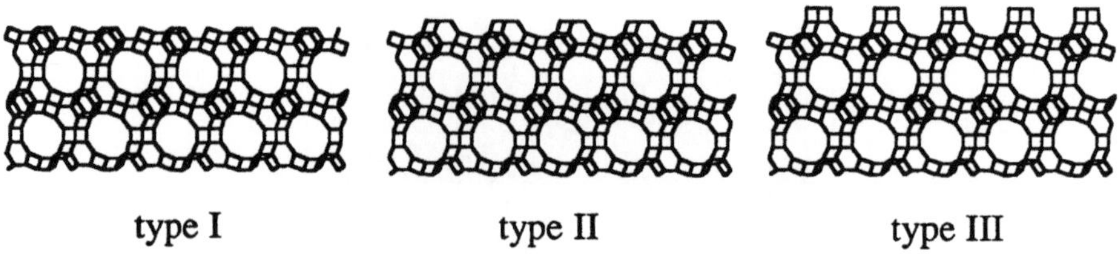

Fig. 15. Three different surface structure models for the $\{111\}_c$ surface of FAU viewed along $[110]_c$

method, fits the simulated image of the type III surface. We have therefore observed two different types of surfaces, that is, structures (I) and (III) in FAU. Both have the common feature that they are related with completed D6R. Structure (II), on the other hand, consists of a S6R at the surface. We propose therefore that the D6R is an important unit in the growth for FAU [11, 12, 17]. This does not necessarily mean that the crystal is formed by the unit of D6R from the gel. Different polymeric molecules may be adsorbed and desorbed in their lifetimes. However once D6R is formed at the surface step, this is so stable that it will not desorb and the step advances to the next. D6R is the key unit for the crystal growth of FAU and Si/Al ordering of the framework. This evidence was also observed inside the crystals as a reflection of growth procedure. The details are discussed in [12].

6.2.3
Clusters on or in Zeolites

There have been many attempts aimed at incorporating metal or metallic compound particles into the spaces of zeolites. The accurate characterization of the particles, which consists of the determination of the particle sizes and their location relative to zeolite framework structures, is the main issue. High resolution TEM is a powerful and direct method for this purpose, but it is not easy to determine decisively whether the particles are located inside the channels of the zeolite or on the external surface of the crystals if the particle size is small. This is because the strong contribution from the framework structure masks the contrast in the HRTEM image from the clusters especially when incident electrons are parallel to the direction of the channels in order to obtain a structure image, and because the electron-optical effect causes an artifact. Furthermore, zeolites are quite electron-beam sensitive as mentioned previously. In order to overcome this difficulty, a few methods have been proposed: image processing [29, 52], HRTEM observation of serial ultra thin sectioned specimens [53, 54] and the Z-contrast method [55, 56]. It is normally believed that the particles are inside, if the particles are not observed on the external surface in the HRTEM image when looking at the direction perpendicular to the channel. A question then arises, how can small particles be observed in HRTEM images under certain conditions? My present answer to this question is given in the next section.

6.2.3.1
Pt/K-LTL

Pt-containing LTL has attracted a lot of attention as a catalyst. The Pt particles are sputtered on the outer surface of the crystals; therefore, we are sure that the particles are on the external surfaces in this experiment. With the electron beam parallel to the channels, we can observe contrast from the Pt clusters which are on the (001) plane, if the particles are larger than approximately 30 Å and this is more than four times larger than the channel aperture. It is easier to observe the particles on the (001) plane by taking HRTEM images with the electron beam perpendicular to the channels.

Figure 16 shows a set of sequential HRTEM images. Platinum particles are stationary on the surface during the electron beam irradiation, which is different

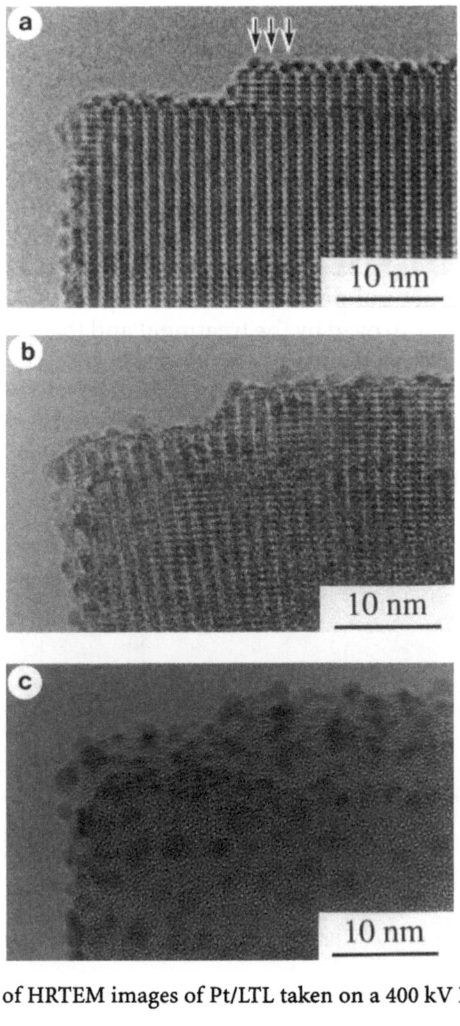

Fig. 16a–c. A sequence of HRTEM images of Pt/LTL taken on a 400 kV EM with [100]

from the previous observation of metallic particles [57]. The Pt particles, with a diameter of ca. 10–15 Å, are situated on the (001) surface. They can be seen situated above the row of channel openings at the surface in the projection shown in Fig. 16a. It is clear from all the observations that 10–15 Å Pt particles are easily detectable. These Pt particles may be useful as arrayed dots although they are materials not confined in the spaces but on the external surfaces. One can see the lattice fringes of Pt particles in both Figs. 16b and c. It is quite obvious that the K-LTL crystal changes its morphology and the Pt particles migrate and change their sizes due to the influence of the beam. Therefore one must be careful to measure the particle size from the HRTEM image after the destruction of the framework, although it is easier to observe contrast from the particles [7, 18].

6.2.3.2
Metal Compounds in FAU

There are several methods to incorporate metal compounds into the spaces of zeolites. Among them, two methods, (i) to incorporate volatile compounds followed by a chemical reaction, and (ii) ion exchange with metal ions followed by reduction and chemical reaction, are the most common procedures. Molybdenum oxide/sulfides and iron oxide clusters were synthesized in the spaces of FAU by a few cycles of chemical reactions of adsorbed $Mo(CO)_6$ and $Fe_2(CO)_5$ with oxygen or H_2S gas. Clusters of MoO_3, MoS_2 and Fe_2O_3 have been successfully fabricated in the spaces of FAU [58–60]. The most important point of this study, from a structural point of view, is the confirmation that the framework structure is not destroyed by the treatment and that no surface precipitates are observed on the external surfaces. An example of a HRTEM image of MoS_x/FAU (Fig. 17) clearly shows molybdenum sulfide, perhaps MoS_2, clusters are incorporated into the spaces without any serious damage to the framework structure. A variation of contrast, which may correspond to different sizes of the clusters, is observed.

Fig. 17. An HRTEM image of MoS_2/FAU taken on a 400 kV EM with $[110]_c$

6.2.4 Modulation and Modification of the Framework Structure

6.2.4.1 Si/Al Concentration Modulation in MOR

Modulation of the Si/Al ratio is common in a zeolite framework, and this is normally observed through an X-ray analysis (compositional mapping) in SEM. In order to observe the modulation by HRTEM, a fine probe is needed to detect it. Fortunately, the Al atoms in the framework always carry the same number of cations and produce electric fields with multipoles, which give rise to physical adsorption. If adsorbed atoms have large enough scattering amplitudes to produce contrast in HRTEM images, then we can detect the modulation on an atomic scale through the contrast.

During the study of making isolated Se-chains in the one-dimensional channels of synthetic MOR, we observed an unusual black and white band contrast in the HRTEM image (Fig. 18). This peculiar contrast was determined as an amplitude contrast due to modulation of the Se-content along the b axis [20,21]. It has been confirmed from our optical absorption spectra of Se confined in the spaces that a finite number, 10–20, atoms of Se are physically adsorbed at each dipole [22]. Selenium atoms have large electron-scattering amplitude; therefore, they can be used as probes to detect the compositional variation of Al in the framework on the unit cell size by producing contrast in the HRTEM images. It should be mentioned that the crystal of MOR used had a plate-like character, extended along the b-axis and thin along the c-axis. This is quite different from natural MOR or other synthetic MOR single crystals which are long along the c-axis. Compositional modulation may be responsible for this characteristic shape, through a dependence of the crystal growth rate (anisotropic) on the Si/Al ratio under the synthetic conditions.

Fig. 18. An HRTEM image of Se-MOR taken on a 1000 kV EM with [001]

6.2.4.2
Dealumination of the Framework

As a modification of the framework, dealumination is a well-established technique for improving thermal stability and enhancing acidity of zeolites. We were able to remove quite a number of Al atoms from the framework using this treatment. In order to observe the effect of dealumination on an atomic scale, especially on the external surfaces, the early stage of the process should be studied by HRTEM [56]. The very regular octahedral shape of FAU and hexagonal plate shape of EMT were chosen and mildly dealuminated by hexafluorosilicate. Figure 19 shows HRTEM images of FAU after dealumination taken with ⟨110⟩. Figure 19b is a high magnification image from the top part of the same crystal shown in Fig. 19a. The crystal shape of FAU is an octahedron and therefore the projection along the direction appears as a rhombus. The prominent features for dealuminated FAU observed by HRTEM are summarized as follows: (i) a dense non-crystalline layer, which follows the original crystal shape, of approximately 20-Å thickness is formed, (ii) between this and the crystalline phase is a layer of less dense amorphous material approximately 80-Å thick, (iii) many mesopores are formed between the amorphous layers and crystals, (iv) mesopores are bounded on all sides by $\{111\}_c$ surfaces, and (v) defects are preferentially attacked. The suggestion that the dense outer layer is deposited silica [61] has been confirmed by recent work [unpublished]. An HRTEM image of EMT

a b

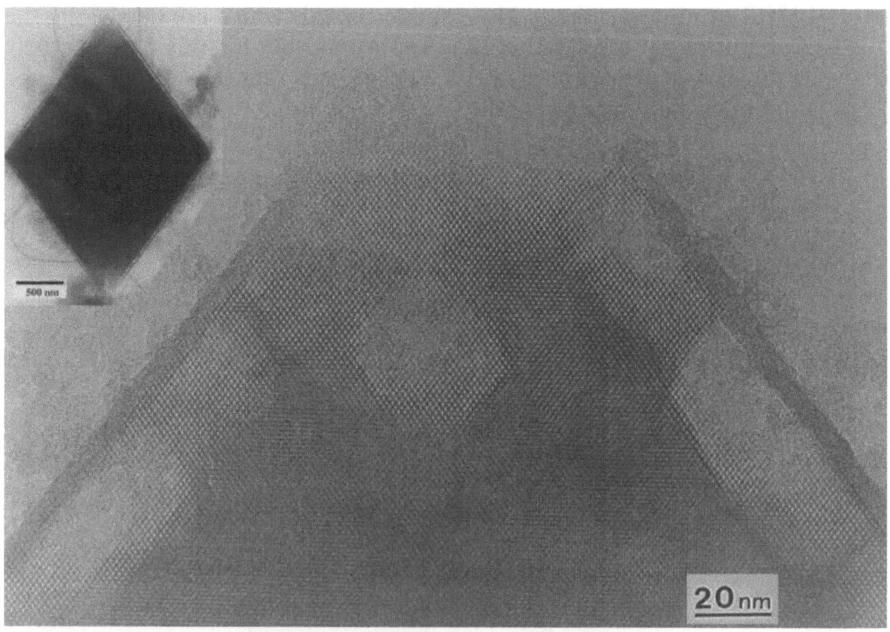

Fig. 19 a, b. An HRTEM image of FAU dealuminated by ammonium hexafluorosilicate taken on a 400 kV EM with [110]

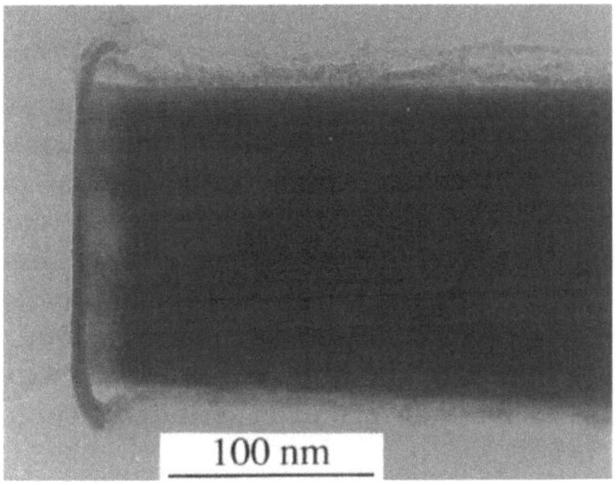

Fig. 20. An HRTEM image of EMT dealuminated by ammonium hexafluorosilicate taken on a 400 kV EM with [100]

viewed perpendicular to the c axis ([100]) after dealumination is shown in Fig. 20. The feature due to the dealumination of EMT looks different but the essential point is the same for both FAU and EMT. From the images it may be speculated as a summary that the apparent change by dealumination is the reverse of the crystal growth process. In the case of the intergrown samples, dealumination occurs preferentially in regions with a higher level of stacking disorder and for applications where some mesoporosity is desirable, this is potentially a method of producing a controlled number of larger pores.

6.2.5
New Structure Determination

6.2.5.1
BEA

Zeolite beta (BEA) was synthesized in 1967 and showed high catalytic activities, from which it was speculated that BEA has a three-dimensional 12-ring framework structure. But the structure was not solved. This is because crystals of BEA always contain severe structural faulting and hence show strong diffuse scattering in diffraction patterns. In 1988, Treacy and Newsam succeeded in solving the structures by use of primarily electron microscopy [13]. The tertiary building unit (TBU), a constituent of the rod, and its connectivity relation were determined from HRTEM images and ED patterns. They proposed structure models with an interpenetrating channel arrangement so as to keep high sorption capacity despite the presence of extensive stacking disorder. Two polymorphs, A and B, which are described by a different stacking sequence of the sheets, are proposed as the end members. The basic process in solving the structures of BEA and ETS-10 is very similar.

6.2.5.2
SSZ-26 and SSZ-33

The new zeolites SSZ-26 and SSZ-33 were synthesized purposefully to contain a three-dimensional pore system comprised of intersecting 10- and 12-ring pores [62]. Their basic structural components, such as pore diameter and channel dimensions, were proposed from independent measurements of N_2 and 2,2-dimethylbutane adsorption, and density measurement. The unit cell with a = 13.26 Å, b = 12.33 Å, and c = 21.08 Å (orthorhombic) for SSZ-33 was determined by X-ray powder diffraction. The essential points of the framework structures were determined by the observations of ED patterns and HRTEM images. But it seems to me that further experimental confirmation might be necessary.

6.2.5.3
ETS-10

A new microporous titanosilicate ETS-10 was first synthesized in 1989 [63], but the structure remained unsolved until recently [14–16]. The process to determine the framework structures for BEA, SSZ-26, SSZ-33 and ETS-10 is similar, so the case for ETS-10 is explained here in detail. The SEM image (Fig. 21) suggests that the material has a pseudo 4-fold symmetry along an axis indicated by the large arrow (hereafter this direction is referred to as the z-axis), and also reveals faults perpendicular to the z-axis indicated by the small arrows. This is clearly confirmed from HRTEM images and ED patterns (see [14–16]). The basic unit cell was determined by ED patterns to be triclinic with a = b = 15.1 Å, c = 14.9 Å, $\alpha = \beta = 104.7°$ and $\gamma = 90°$.

High-resolution TEM images and corresponding ED patterns taken along an axis perpendicular to z (assigned as the x-axis) are shown in Fig. 22 together with a schematic drawing. In the images a number of characteristic features are seen:

- Large white dots (indicated by white stars) are arranged along the y-axis with a period of ca. 15 Å. There is a group of black contrast between the two white dots as indicated by black stars. These, as a whole, form sheets, three of which are indicated by the medium-sized arrows (Fig. 22a, b).

Fig. 21. An SEM image of ETS-10

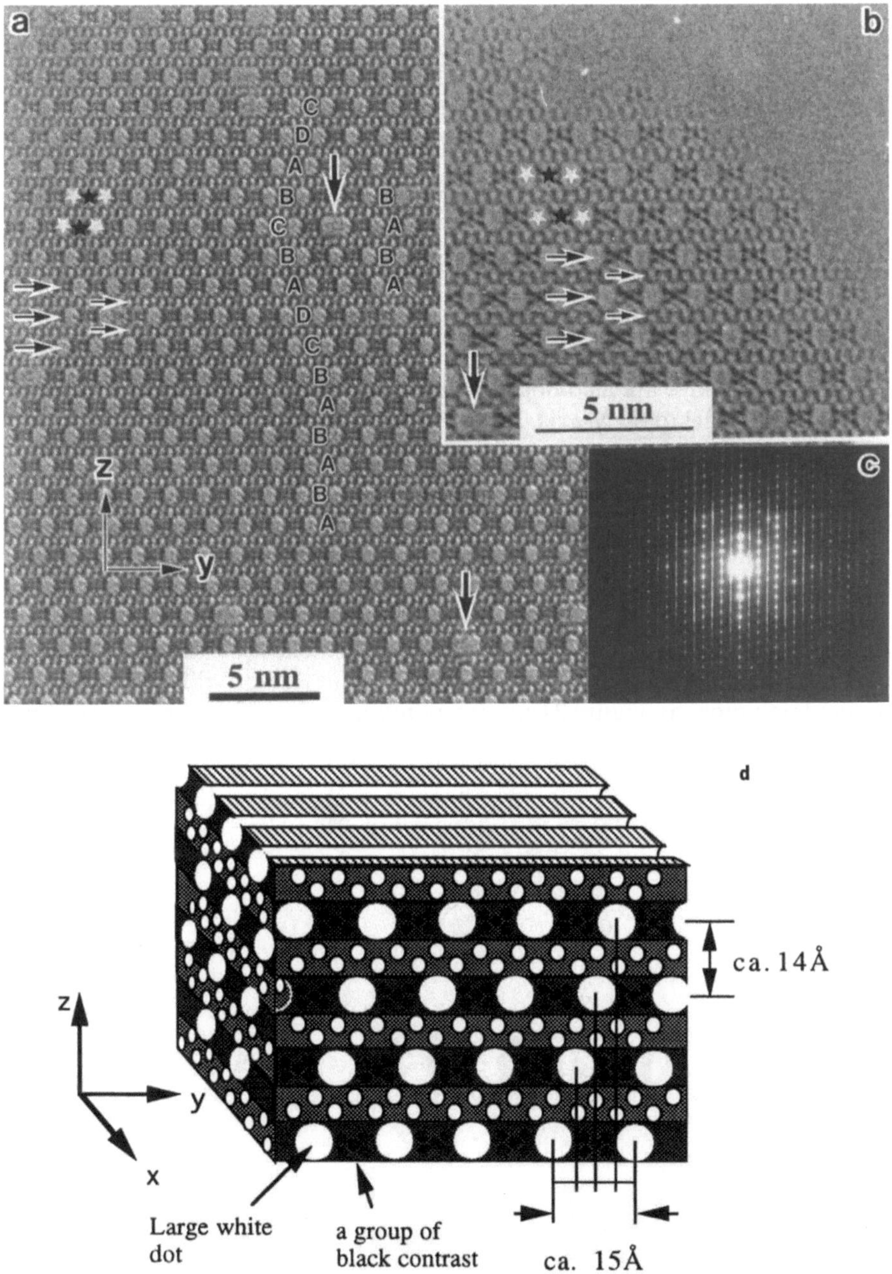

Fig. 22 a–d. High resolution TEM images of ETS-10 along the x-(or y-) axis taken on a 400 kV EM (**a** and **b**), the corresponding ED pattern (**c**) and a schematic drawing of HRTEM images (**d**)

- Between the successive sheets, rows of small white dots are arranged in a zigzag manner along the y-axis as indicated by small arrows (Fig. 22a, b).
- Large white dots in successive sheets shift by one quarter of the period of the dots along the y-axis (ca. 15 Å) either to the left or the right, and the structure can then be described by the stacking of the sheets in the z-direction using ABCD or ABAB, etc (Fig. 22a, d).
- There are many large channels, indicated by large arrows, in the sheets created by stacking faults. The stacking sequence of the left side of the larger channel is ABCBADC and that of right side is ABAB as marked in the image (Fig. 22a, b).
- Diffuse streaks with some maxima are therefore observed in the ED pattern along the direction perpendicular to the sheets (Fig. 22c). Exactly the same features as above are also observed in an HRTEM image taken with the incidence parallel to the y-axis.

From the above observations, a structure unit (a rod) and its connectivity relationship are determined as shown in Fig. 23, where Si, O and Ti are distinguished by the different solid circles, by considering the framework composition ratio of Si/Ti = 5. The rod consists of a one-dimensional titanium oxide chain $(\text{-O-Ti-O-Ti-O-})_n$ and two 3-membered rings with Si on either side.

Many different stacking sequences are observed in HRTEM images. The only rule is that a shift must be introduced to the next sheet by one quarter of the period of the dots. Figure 24 shows two extremes of ETS-10 which represent two ordered polymorphs. Figure 24a, corresponding to an ABAB... repeat stacking,

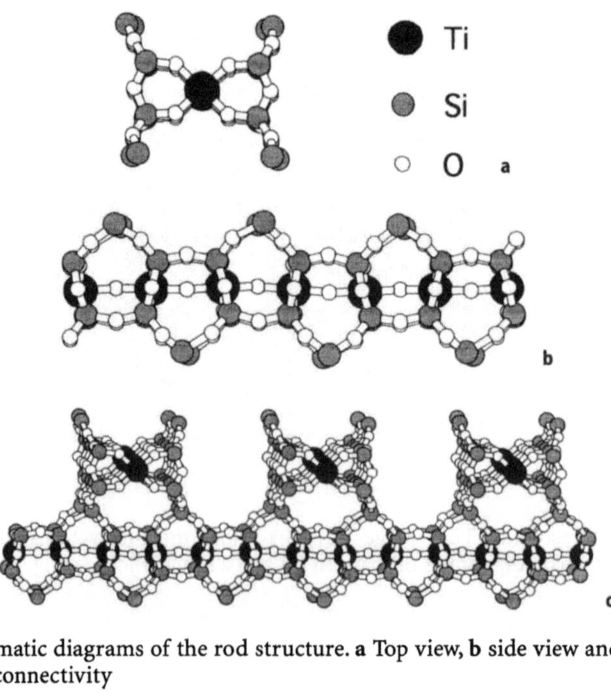

Fig. 23a–c. Schematic diagrams of the rod structure. **a** Top view, **b** side view and **c** a possible example of the connectivity

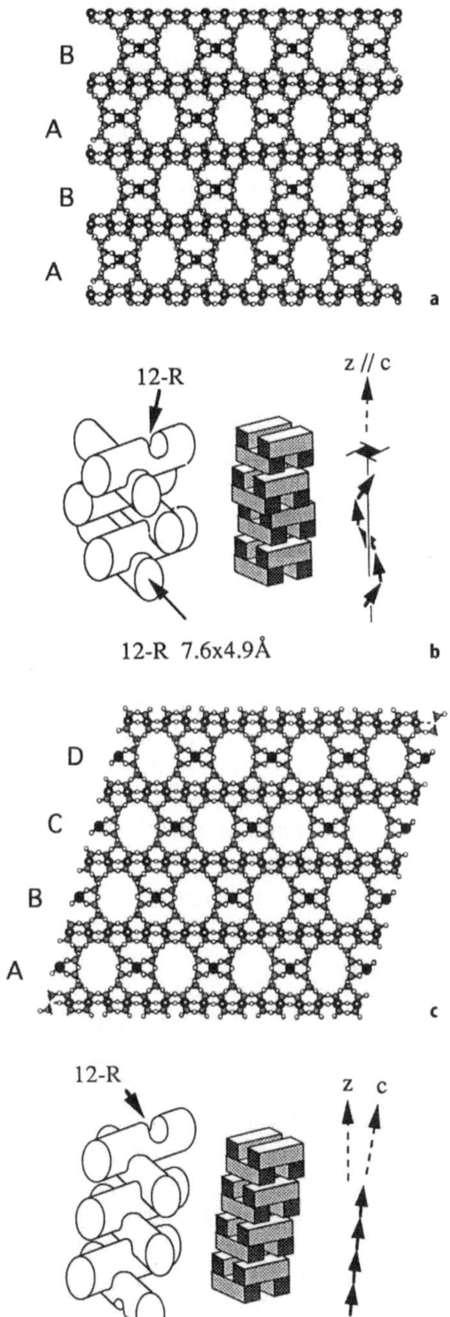

Fig. 24a–d. Projected models for the structures of two end members, **a** polymorph A (ABAB stacking) and **c** polymorph B (ABCD stacking). Schematic channel systems for polymorph A (**b**) and for polymorph B (**d**)

has unit cell parameters a = b = 14.85 Å and c = 27.08 Å (the tetragonal space-group $P4_122$ or $P4_322$). There is a spiral channel with a 12-membered ring aperture which spirals along the c-axis in a clockwise or anti-clockwise direction as shown in Fig. 24b, and this is a chiral crystal system. Figure 24c, corresponding to an ABCD... repeat stacking, has unit cell parameters a = b = 21.00 Å, c = 14.51 Å and $\beta = 111.12°$ (the monoclinic space-group C2/c). In this polymorph a straight 12-membered ring channel runs parallel to the c-axis as shown in Fig. 24d.

6.2.5.4
New Porous Materials, MCM and FSM

Electron microscopy also plays an important role in characterizing porous materials [64–69]. In these materials, the following structural points have to be characterized: i) whether the wall is amorphous or crystalline and, if it is amorphous, how to specify it, ii) wall thickness, iii) pore diameter, and iv) an arrangement of pores including the "symmetry" of the system. Electron diffraction patterns can be obtained from a "single domain" and therefore they give information on the problems i) and iv), and HRTEM images give information on ii), iii) and iv). Electron microscopy will shed light for finding a difference between MCM-41 [64 – 67] and FSM-16 [69], if they are different. Recent work by Alfredsson and Anderson showed interesting results on the structure of cubic MCM-48 using electron microscopy with computer modeling and simulation [68]. This kind of technique, i.e. a combination of EM and other computing methods such as MD or Monte Carlo simulation, will extend the study of the fine structures of zeolites with some kind of disorder or irregularity for which classical diffraction techniques cannot give enough information. Electron crystallography (Sect. 8.3) will open a new field to characterize/determine structures of new porous materials.

7
Electron Radiation Damage

Zeolite materials are electron beam sensitive, and Bursill and Thomas have discussed the damage process and methods for stabilization [39–41]. There are two major elementary processes of electron damage; knock-on and radiolytic damage. The first arises from direct interactions between incident electrons and the nucleus of the atoms in the specimen. The cross-section for the knock-on process increases from a certain threshold energy, for moving an atom from the proper site into an unstable position, with increasing accelerating voltage. The second consists of several electron excitations, such as (i) inner shell ionization, (ii) plasmon loss, (iii) creation of locally bound electron-hole pairs (excitons), and the cross-section for radiolysis decreases with increasing accelerating voltage. Csencsits and Gronsky discussed these two elementary processes for FAU, and measured electron dose to vitrification as a function of accelerating voltage and of Si/Al ratio of the frameworks. They confirmed experimentally that FAU materials with higher Si/Al ratios are more stable under electron irradiation, and that the radiolytic process is the main part of vitrification in the range 80–200 kV [70]. It is rather difficult to indicate the electron dose sufficient to destroy the crystallinity of zeolites, as it is strongly dependent on the type of zeolite, Si/Al ratio, water content, energy of incident electrons and so on. Treacy and Newsam claimed a dose rate of about 50 electrons $Å^{-2}$ s^{-1} for LTL [71] and McComb and Howie mentioned a dose from 10^3 to 10^5 electrons $Å^{-2}$ [72].

The radiation damage process is sometimes not uniform, as first observed by Treacy and Newsam for LTL [71]. The sequence of HRTEM images of LTL taken with [100] is shown in Fig. 25 under electron irradiation. From these images it is very clear that the degradation of LTL under an electron beam is not uniform. The crystallinity at the top and bottom (001) surfaces stays longer than at the middle part, and the crystal shape changes from a cylindrical plate to a cotton-reel shape (to form a waist) as shown schematically in Fig. 25d. This may be explained by the amorphous material produced by the electrons having a higher density than crystalline LTL. Treacy and Newsam have also confirmed the previous observation that the rate of vitrification depends sensitively on the degree of hydration.

Although the presence of water in a zeolite crystal especially shortens the lifetime of the crystal under the electron beam, we now realize that the most important factor for stability of the crystal is the quality of their crystallinities. We can observe beautiful HRTEM images even from fully hydrated ordinary Na-FAU (Si/Al = 2.8) [7].

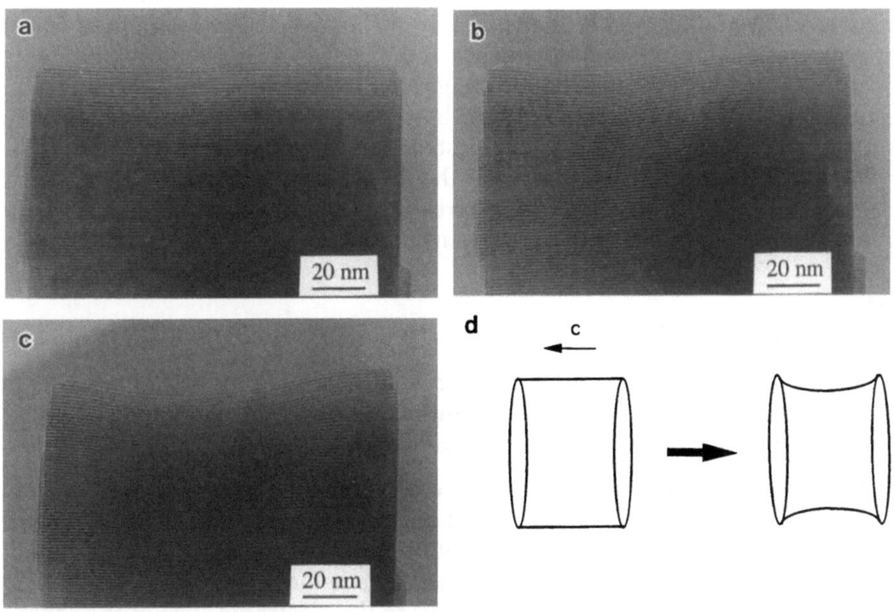

Fig. 25 a–d. A sequence of HRTEM images of LTL taken on a 400 kV EM with [100] in order to show anisotropic damage under an electron beam (**a–c**). **d** Schematic drawing of **c**

8 Other Topics

8.1 Image Processing

To obtain the intuitive HRTEM images, in which the contrast is proportional to the projected potential for electrons as discussed in Sect. 4, it is necessary to include as many reflections for image formation (i.e., Fourier synthesis of diffracted beams) as possible. Therefore, it might be necessary by extending zero cross u_0 to have an EM with better resolution which is essentially determined by the CTF as discussed in Sect. 4.2. But in the case of zeolites, the situation is complicated. Information at small wave vectors, which is most important for image formation in zeolites with a large lattice constant, is inevitably transferred less as the resolution of the EM becomes higher. This causes artificial contrast at the centers of channels in HRTEM images when the images are taken along the channel direction. It is therefore important to distinguish this contrast from that of the materials confined in the spaces of zeolites. By knowing the origin of the artifacts we can remove artificial contrast with image processing [28, 29, 52] as a makeshift.

8.2 High-Voltage HRTEM

The advantages of using high-voltage EM for the study of the fine structures of zeolites are i) better resolution, ii) less damage, iii) observation of the fine structures from thicker specimens, and iv) fulfilment of the weak phase object approximation through the dependence of the interaction parameter σ on E (see Sects. 1.3 and 4.1). The use of high-voltage HRTEM will become more important especially for the observation of grain boundaries in membrane zeolites or in sintered zeolites without a process/destruction. An advantage of using high-voltage HRTEM for an observation of the fine structure from thick specimens has already been shown [7, 52]. We interesting results showing all the above advantages for zeolites by using a new re-installed 1.25 MeV EM at our university.

8.3 Electron Crystallography

In addition to the advantages of electron scattering over X-ray and/or neutron scattering as discussed in Sect. 1.3, I want to mention two other points in particular for the study of fine structures of zeolites and clusters/zeolites.

1. Electron atomic scattering factors are very sensitive to the electronic state (outermost electrons) of the atom at small scattering vectors, and it is also anisotropic in the vectors for covalent bonding character [73]. In the case of zeolites, fortunately we have many reflections in the range where the electron atomic scattering factor is very sensitive to the electronic state of the constituent atoms. For example, the values of sin θ/λ for the 111 reflection of FAU (a = 24.7 Å) and 200 of LTA (a = 12.4 Å) are 0.035 and 0.08 $Å^{-1}$, respectively.

2. Electron atomic scattering factors change with atomic number but the dependence is not as strong as for X-ray scattering (see International Tables for X-ray Crystallography, vol. 4, IUCR). We can obtain a relatively large amount of information from a constituent atom of a light element (small Z number) in clusters which are composed with large Z number atoms.

Recent development in recording media for electrons with a computer enables us to treat the data, ED patterns and HRTEM images, more quantitatively. If we are able to define experimental conditions, such as crystal thickness, beam tilting to the crystallographic axis and focus conditions, we can use intensity distributions of ED patterns in a quantitative way for structure determination by relying on the theory of electron dynamic scattering. Even a relatively low-resolution HRTEM image will be useful for a phase recovery of reflections. Furthermore, a combination of computational methods with qualitative analysis may open a new field for the study of the structures of zeolites and clusters in or on zeolites.

Finally, the recent work by Nicopoulos and his colleagues should be mentioned as the first application of electron crystallography to a study of zeolite structure, in this case, MCM-22 [74].

9
Notes Added at Revision (24 June 1997)

9.1
Mesoporous Materials

Many papers have been published on mesoporous materials to study the points mentioned in Sect. 6.2.5.4 by EM, but most of all HRTEM images of MCM-41 have been used only to show that the channels are hexagonally arranged [64–67]. The images should be treated with care as they contain a lot of information although the images depend on experimental conditions, especially on out of focus and specimen thickness. Recently, Sakamoto et al. observed HRTEM images of FSM-16s, both as-synthesized and calcined [75, 76]. From careful image simulations it is reported that a circular channel can be imaged not as a circle, but as a regular hexagon under certain experimental conditions (Fig. 26). If the crystal thickness increases, this effect becomes prominent under ordinary focus conditions. For as-synthesized FSM-16, they showed the following:

- The wall thickness is uniform and corresponds to double sheets of silicate. Therefore there is no indication of a periodic change of single and double sheet silicates as expected from the proposed growth model by Inagaki et al. [69], the Folded Sheet Mechanism, from which the name of FSM-16 comes.
- The shape of channels is more or less hexagonal.
- The characteristic external morphology of a "hexagonal plate" (angles of $2\pi/3$ between two neighboring edges in projection), which is observed in MCM-41, is not observed.

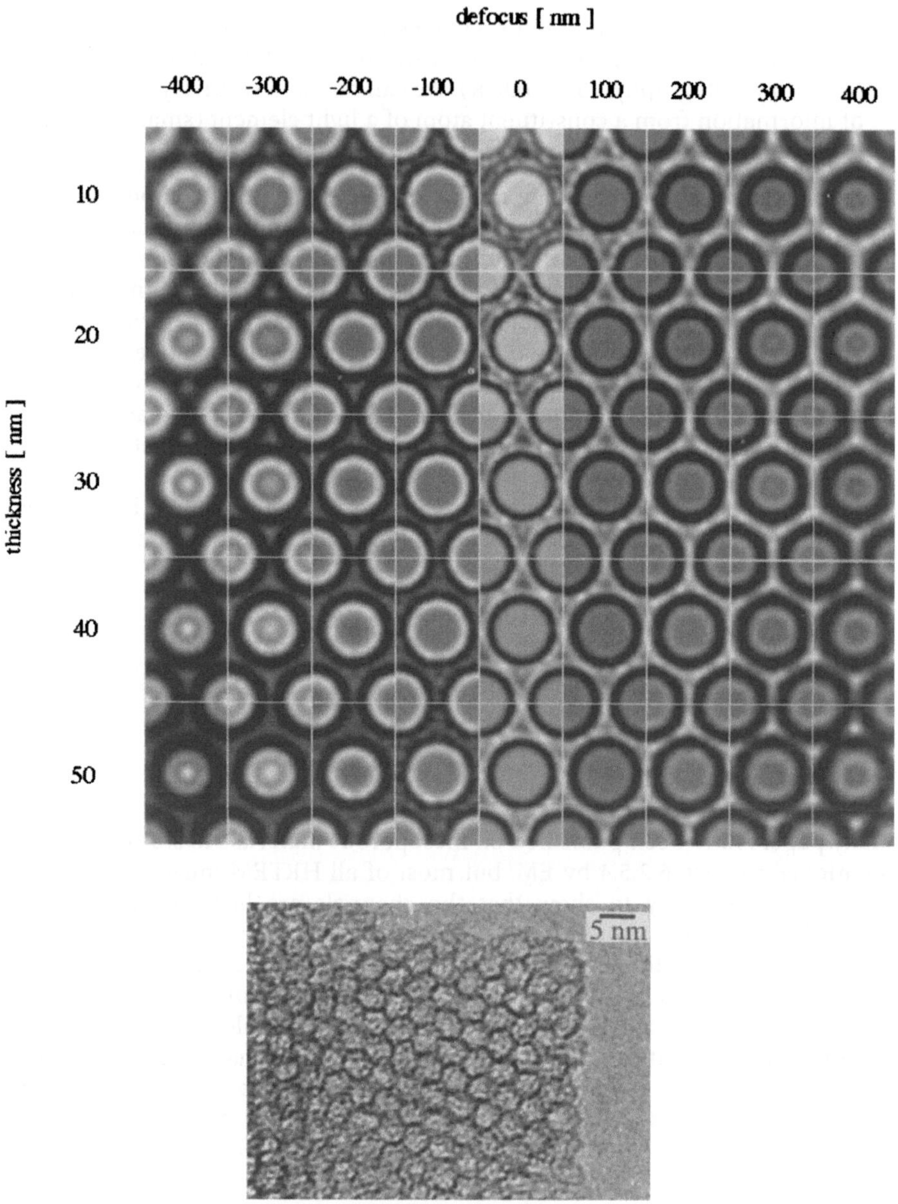

Fig. 26. A simulated image of FSM-16 with a circular aperture of the channel (*above*), and an observed HRTEM image of as-synthesized FSM-16 (*below*)

- The length along the channel direction is much longer than that perpendicular to the direction.

They proposed a growth mechanism of FSM-16 based on their observations. They also reported the change caused by a process of calcination, that is, the wall becomes much thicker and the channel shape changes from hexagonal to circular after the process, which is consistent with the previous observation in X-ray diffraction [75].

9.2 Quantitative Analysis of ED Patterns and HRTEM Images

New improvements in recording media, such as the imaging plate and the slow-scan CCD camera with an output of higher linearity and a larger dynamic range than ordinary photographic films, enable us to analyze ED patterns and HRTEM images quantitatively. An example has been shown recently for ZSM-5 to reveal TPA cations, which are used as templates for synthesis. The presence of TPA is shown in a Fourier map of electrostatic potential distribution of ZSM-5 with TPA (Fig. 27), which was obtained from the ED pattern of [010] incidence [77]. The authors discussed the amount of TPA in the channels from the observed contrast. Generally, to obtain a Fourier map, it is necessary to speculate a relative phase relationship between the reflections, but from the corresponding HRTEM image we can determine experimentally the relationship for the reflections

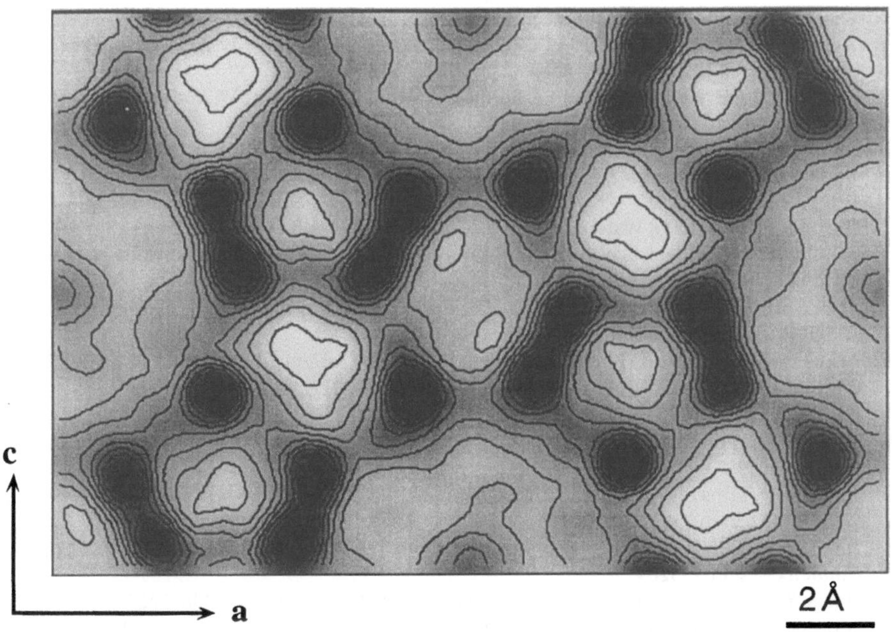

Fig. 27. A Fourier diagram obtained from 98 independent reflections in the ED pattern of TPA/ZSM-5, [010] incidence

within the window of CTF (smaller than u_0, as discussed in Sect. 4.1). A study to apply the direct method, which has been developed for X-ray diffraction to analyze ED patterns and/or to combine analysis of both ED patterns and HRTEM images for clarifying structures of clusters confined in the spaces of zeolites, is underway.

Acknowledgements. The author is grateful to D. Watanabe, S. Andersson, J.M. Thomas, the late J.V. Sanders, T. Ohsuna, J-O Bovin, V. Alfredsson, M.W. Anderson, K. Hiraga, N. Ohnishi and Y. Sakamoto for their continuous encouragement, support and collaboration. Thanks are also expressed to the Royal Society and the British Council (UK), the Swedish Natural Science Research Council, NFR, CREST, the Japanese Science and Technology Corporation (JST), and Ministry of Education, Science and Culture, Japan for their support.

References

1. Bush H (1926) Ann Phys Lpz 81:974
2. Knoll M, Ruska E (1932) Z Physik 78:318; Ruska E (1980) The early development of electron lenses and electron microscopy. S Hirzel Verlag, Stuttgart
3. Breck DW (1974) Zeolite molecular sieves. Wiley, New York
4. Terasaki O, Thomas JM, Millward GR (1984) Proc R Soc Lond A395:153
5. Terasaki O (1991) Acta Chem Scand 45:785
6. Terasaki O (1994) J Electron Microsc 43:337
7. Terasaki O, Ohsuna T (1995) Catal Today 23:201
8. Millward GR, Ramdas S, Thomas JM (1985) Proc R Soc A399:57
9. Millward GR, Thomas JM, Terasaki O, Watanabe D (1986) Zeolites 6:91
10. Anderson MW, Pachis KS, Prebin F, Carr SW, Terasaki O, Ohsuna T, Alfredsson V (1991) J Chem Soc Chem Commun:1660
11. Terasaki O, Ohsuna T, Alfredsson V, Bovin J-O, Watanabe D, Carr SW, Anderson MW (1993) Chem Mater 5:452
12. Ohsuna T, Terasaki O, Alfredsson V, Bovin J-O, Watanabe D, Carr SW, Anderson MW (1996) Proc R Soc Lond A452:715
13. Newsam JM, Treacy MMJ, Koetsier WT, Gruyter CB (1988) Proc R Soc Lond A420:375
14. Anderson MW, Terasaki O, Ohsuna T, Philippou A, MacKay SP, Ferreira A, Rocha J, Lidin S (1994) Nature 367:347
15. Ohsuna T, Terasaki O, Watanabe D, Anderson MW, Lidin S (1994) In: Weitkamp J, Karge HG, Pfeifer H, Hölderich W (eds) Zeolites and related microporous materials: state of the art 1994. Elsevier, Amsterdam, p 413
16. Anderson MW, Terasaki O, Ohsuna T, Philippou A, MacKay SP, Ferreira A, Rocha J, Lidin S (1995) Phil Mag 71B:813
17. Alfredsson V, Ohsuna T, Terasaki O, Bovin J-O (1993) Angew Chem Int Ed Engl 32:1210
18. Ohsuna T, Terasaki O, Hiraga K (1996) Mat Sci Eng A217/218:135
19. Bogomolov VN (1978) Sov Phys Usp 21:77
20. Terasaki O, Yamazaki K, Thomas JM, Ohsuna T, Watanabe D, Sanders JV, Bary JC (1987) Nature 330:58
21. Terasaki O, Yamazaki K, Thomas JM, Ohsuna T, Watanabe D, Sanders JV, Bary JC (1988) J Solid State Chem 77:72
22. Nozue Y, Kodaira T, Terasaki O, Yamazaki K, Goto T, Watanabe D, Thomas JM (1990) J Phys Condens Matter 2:5209
23. Nozue Y, Kodaira T, Ohwashi S, Goto T, Terasaki O (1993) Phys Rev B48:12253
24. Terasaki O, Tang ZK, Nozue Y, Goto T (1991) Mat Res Soc Symp Proc 233:139
25. Terasaki O (1993) J Solid State Chem 106:190
26. Bursill LA, Lodge EA, Thomas JM (1980) Nature 286:111; J Catal 103:466

28. Alfredsson V, Terasaki O, Bovin J-O (1990) J Solid State Chem 84:171
29. Alfredsson V, Terasaki O, Bovin J-O (1993) J Solid State Chem 105:223
30. Menter J W (1956) Proc R Soc Lond A236:119
31. Menter JW (1958) Adv Phys 7:299
32. Kerr IS, Gard JA, Barrer RM, Galabova IM (1970) Am Mineral 55:441
33. Gard JA, Tait JM (1971) Adv Chem Series 101:230
34. Kokotailo GT, Sawruk S, Lawton SL (1972) Am Mineral 57:439
35. Frety R, Ballivet D, Barthomeuf D, Trambouze Y (1972) CR Acad Sci Paris, Serie C t275:1215
36. Allpress JG, Sanders JV (1973) J Appl Cryst 6:165
37. O'Keefe MA, Sanders JV (1976) Optik 46:421
38. Sanders JV (1978) The resolution of electron microscopes, physics of materials. University of Melbourne Press, p 244
39. Bursill LA, Lodge EA, Thomas JM (1980) Nature 286:111
40. Bursill LA, Thomas JM (1981) J Phys Chem 85:3007
41. Bursill LA, Thomas JM, Rao KJ (1981) Nature 289:157
42. Tschernich RW (1992) Zeolites of the world. Geoscience
43. Born M, Wolf E (1984) Principles of optics, 6th edn. Pergamon, Oxford, p 424
44. Terasaki O, Ohsuna T, Watanabe D, Kessler H, Schott-Darie C (1995) Stud Surf Sci Catal 98:50
45. Delprato E, Delmotte L, Guth JL, Huve L (1990) Zeolites 10:546
46. Kokotailo GT, Lawton SL, Olson DH, Meier WM (1978) Nature 272:437
47. Kokotailo GT, Chu P, Lawton SL, Meier WM (1978) Nature 275:119
48. Terasaki O, Ohsuna T, Sakuma H, Watanabe D, Nakagawa Y, Medrud RC (1996) Chem Materials 8:463
49. Fyfe CA, Feng Y, Grondey H, Kokotailo GT, Mar A (1991) J Phys Chem 95:3747, and refs cited therein
50. Millward GR, Ramdas S, Thomas JM, Barlow MT (1983) J Chem Soc Faraday Trans 2 79:1075
51. Terasaki O, Thomas JM, Millward GR, Watanabe D (1989) Chem Materials 1:158
52. Terasaki O, Ohsuna T, Alfredsson V, Bovin J-O, Watanabe D, Tsuno K (1991) Ultramicroscopy 39:238
53. Alfredsson V, Terasaki O, Blum Z, Bovin J-O, Karlsson G (1995) Zeolites 15:111
54. Bovin J-O, Alfredsson V, Karlsson G, Carlsson A, Blum Z, Terasaki O (1996) Ultramicroscopy 62:277
55. Treacy MMJ, Howie A, Wilson CJ (1978) Phil Mag A38:569
56. Roce SB, Koo JY, Disco MM, Treacy MMJ (1990) Ultramicroscopy 34:108
57. Wallenberg R, Bovin J-O, Smith DJ (1985) Naturwissenschaften 72:S539
58. Okamoto Y, Kikuta H, Ohto Y, Nasu S, Terasaki O (1997) Stud Surf Sci Catal 105:2051
59. Okamoto Y, Katsuyama H, Yoshida K, Nakai K, Matsuo M, Sakamoto Y, Yu J, Terasaki O (1996) J Chem Soc Faraday Trans 92:4647
60. Sakamoto Y, Togashi N, Terasaki O, Ohsuna T, Okamoto Y, Hiraga K (1996) Mat Sci Eng A217/218:147
61. Ohsuna T, Terasaki O, Watanabe D, Anderson MW, Carr SW (1994) Chem Mater 6:2201
62. Lobo RF, Pan M, Chan I, Medrud RC, Zones SI, Crozier PA, Davis ME (1994) J Phys Chem 98:12040
63. Kuznicki SM, Thrush KA, Allen FM, Levine SM, Hamil MM, Hayhurst DT, Mansour M (1992) In: Occelli ML, Robson H (eds) Molecular sieves. van Nostrand Reinhold, New York, p 427
64. Kresge CT, Leonowicz ME, Roth WJ, Vartuli JC, Beck JS (1992) Nature 359:710
65. Beck JS, Vartuli JC, Roth WJ, Leonowicz ME, Kresge CT, Schmitt KD, Chu CTW, Olson DH, Sheppard EW, McCullen SB, Higgins JB, Schlenker JL (1994) J Am Chem Soc 114: 10834
66. Ciesla U, Demuth D, Leon R, Petroff P, Stucky GD, Unger KK, Schüth F (1994) J Chem Soc Chem Commun 1387

67. Alfredsson V, Keung M, Monnier A, Stucky GD, Unger KK, Schüth F (1994) J Chem Soc Chem Commun 921
68. Alfredsson V, Anderson MW (1996) Chem Materials 8:1141
69. Inagaki S, Fukushima Y, Kuroda K (1993) J Chem Soc Chem Commun 680
70. Csencsits R, Gronsky R (1987) Ultramicroscopy 23:421
71. Treacy MMJ, Newsam JM (1994) Ultramicroscopy 23:411
72. McComb DW, Howie A (1990) Ultramicroscopy 34:84
73. Watanabe D, Terasaki O (1972) NBS Special Publication 364:155
74. Nikopoulos S, González-Calbet JM, Vallet-Regí M, Corma A, Corell C, Guil JM, Pariente JP(1995) J Am Chem Soc 117:8947
75. Inagaki S, Sakamoto Y, Fukushima Y, Terasaki O (1996) Chem Materials 8:2089
76. Sakamoto Y, Inagaki S, Ohsuna T, Ohnish N, Nozue Y, Terasaki O (1998) Microporous and Mesoporous Materials 21:589
77. Ohnishi N, Hiraga K (1996) J Electron Microsc 45:85

Structural Distortions and Modulations in Microporous Materials

W. Depmeier

Universität Kiel, Institut für Geowissenschaften, Olshausenstr. 40, D-24098 Kiel, Germany;
e-mail: depmeier@min.uni-kiel.de

List of Abbreviations 113

1 Introduction 114

2 Ideal and Real Structures in Zeolites 114

3 Zeolite Frameworks 115

4 Bond Lengths and Angles 116

5 Symmetry 118

6 Systematic Tetrahedron Distortion 119

7 Conformational Changes of Zeolite Frameworks 122

8 Structural Relaxation Around Isolated AlO_4 Tetrahedra in a Silica Matrix 126

9 Thermal Expansion 126

10 Structural Modulations 128

11 Rigid Unit Modes 135

12 Conclusions 136

References 137

List of Abbreviations

CAW	Aluminate sodalite of composition $Ca_8[Al_{12}O_{24}](WO_4)_2$
CAM	Aluminate sodalite of composition $Ca_8[Al_{12}O_{24}](MoO_4)_2$
CAS	Aluminate sodalite of composition $Ca_8[Al_{12}O_{24}](SO_4)_2$
SAM	Aluminate sodalite of composition $Sr_8[Al_{12}O_{24}](MoO_4)_2$
SAW	Aluminate sodalite of composition $Sr_8[Al_{12}O_{24}](WO_4)_2$
OTO, TOT	Angles O–T–O, T–O–T

Molecular Sieves, Vol. 2

4R, 6R, 8R	Ring of four, six or eight tetrahedra in the zeolite framework
NMR	Nuclear magnetic resonance
MAS-NMR	Magic angle spinning nuclear magnetic resonance
DOR-NMR	Double rotation nuclear magnetic resonance
QCC	Quadrupole coupling constant
FCD	Framework contribution to the tetrahedron distortion
CCD	Cation contribution to the tetrahedron distortion

1 Introduction

This article will discuss some of the characteristics of the real structures of zeolites and clathrasils. For simplicity, the term "zeolite" will be used for all classes of microporous materials. The term "real structure" indicates that the structure under discussion is distorted from an ideal one, for example, by changing temperature, pressure, different chemical composition, a chemical reaction, or adsorption/desorption processes. The distortion will mainly concern the tetrahedra and the framework conformation. The topology of the framework does not change but also contributes to the distortion.

The article will begin with some necessary remarks on the basic properties of zeolite frameworks, including the geometry and symmetry aspects. Systematic tetrahedron distortion can be induced by the topology of the framework. This will be discussed in Sect. 6. Cooperative movements of the tetrahedra change the conformation of the framework and this can result in structural collapse. The correlation of the cooperative movements can extend to infinity, or can be localized. Examples of both types are discussed in Sect. 7. Structural relaxation of the framework around an isolated defect is the subject of Sect. 8. The action of one intensive variable, viz. temperature, on the framework will be briefly described in Sect. 9. Owing to space limitations the important structural distortions which occur under the action of pressure, during hydration/dehydration processes, or upon adsorption of molecules are not covered in this article. The important subject of structural phase transitions has also been omitted. Modulations are long-period distortions of a structure. The commensurate and incommensurate modulations occurring in the sodalite family will be discussed in some detail. The question will be posed as to whether such modulations can also be expected in large-pore zeolites. Finally, short reference to "Rigid Unit Modes" is made.

2 Ideal and Real Structures in Zeolites

The terms ideal and real structures are used in a similar way to Megaw's "aristotype" and "hettotypes" [1]. The ideal structure is the simplest and most symmetrical member of a structural family. The real structures are more complex, e.g., in their chemical composition, and are often of lower symmetry. For a given zeolite there is only one unique ideal structure, but usually a manifold of real structure types exists. The term "ideal" will also include the meaning of the fully expanded framework in the case of collapsible zeolites. In many cases real structures of zeolites have a partially collapsed framework.

The members of a structural family are related by one-to-one mapping of all their atoms. Zeolites are capable of hosting in the cavities of their frameworks important quantities of non-framework species, i.e., various ions, molecules, or water. Such derivatives will also be regarded as real structures of a given structural family. Stacking variants of different framework types and lattice defects, such as screw or edge dislocations, are beyond the scope of this chapter.

The differences between a given ideal structure and any of its associated low-symmetry relatives in terms of atomic coordinates are so small that their experimental determination is often difficult. If the electron density of the ideal structure is ϱ_0, and the small deviation of any real structure from the ideal structure is designated by $\Delta\varrho$, then the real structure is:

$$\varrho = \varrho_0 + \Delta\varrho \tag{1}$$

In diffraction methods for structural analysis a small $\Delta\varrho$ results in only minor changes in reciprocal space. The contribution of $\Delta\varrho$ to the total diffracted intensity is often of the order of only 1 %, or less. Hence, it is expected that contributions from $\Delta\varrho$ will cause only slight modifications in the ideal diffraction pattern, such as small shifts of peak positions, split powder diffraction lines, weak extra reflections, or small intensity differences. The differences are often of the same order of magnitude as the resolution of the experiment, or even smaller. Without proper care being exercised these differences can easily be missed. This is true for all pseudo-symmetric structures, e.g., perovskites, but zeolites present the additional difficulty that their crystal quality is often low and their size is small. These facts make diffraction experiments on zeolites even more difficult, and their results sometimes questionable.

There are cases where researchers are content just to establish the general building principles of a new zeolite, for example, if they are mainly interested in aspects of its technical applications. Then the knowledge of the idealized structure may be sufficient, and the often laborious task of working out the real structure can be avoided. On the other hand, if it is necessary to know structural details, such as the distribution of cations or the correct symmetry and the associated crystal properties, then a knowledge of the real structure is indispensable [2]. Given their potential in modern applications, the determination, understanding and control of their real structures will become a key issue in the field of zeolites.

3
Zeolite Frameworks

The basic structural building unit of the zeolites is the TO_4 tetrahedron, where T represents a small, highly charged cation, usually Si^{4+} or Al^{3+}. The quite rigid TO_4 tetrahedra are all-corner-connected via common oxygens, thus forming a three-dimensional framework. The framework can be mapped onto a class of mathematical objects, so-called four-connected nets. These and other nets have been studied extensively in the past, not least because they allow the enumeration of possible zeolite frameworks [3–6]. An infinite number of such nets exsists, but only a limited number can be realized as zeolite structures. These nets allow for "reasonable" inter-atomic distances and angles, given that the nodes are not mathematical points, but more or less rigid TO_4 tetrahedra. Klein used these constraints, and the fact that in real structures the number of topologically different TO_4 tetrahedra tends to be small, to develop a graph-based method which provides a complete enumeration of already known and hypothetical new zeolite frameworks for given numbers of independent tetrahedra [7, 8].

If a synthetic or natural zeolite containing a new framework is discovered, it is labelled by a three-letter code, after approval by the Structure Commission of the International Zeolite Association. The code then represents the framework type of the zeolite, regardless of its actual chemical composition, its symmetry, the degree of structural distortion, or the presence or absence of possible non-framework species. In other words, it corresponds to the ideal structure. The framework type code will be used in this chapter where appropriate.

A total of 108 different framework types are listed in the fourth edition of the Atlas of Zeolite Structure Types [9], together with much useful information. A website can be found under http://www.iza-sc.ethz.ch. The information given there includes topological characteristics of the framework, such as the coordinates of the T atoms in the topological symmetry, the loop configuration of the T atoms, and coordination sequences. The number, orientation, size and variability of channels and cavities of a zeolite are important for appraising possible applications. Approximate cell parameters and the channel and cavity diameters are also listed. These results were obtained by a distance-least-squares method [10] assuming a pure SiO_2 framework. Schematic drawings of the respective framework type complete the information. Finally, synthetic or naturally occurring representatives of the framework type are mentioned.

As a very useful tool it is possible to calculate and display the powder pattern for many zeolites. In some cases it is even possible to perform the calculation for several real representatives of a given framework topology, and to study the effects of the change of a real structure on the diffraction pattern.

4 Bond Lengths and Angles

For a given T cation the TO_4 tetrahedron has only a narrow range of variability with respect to its T–O distances and OTO angles, even when it is part of a different framework type. This fact has been demonstrated for many examples, and explained by the partial covalent character of the T–O bond [11–14].

With a new large and highly complete data base on zeolites [15] becoming operational soon, it can be expected that more detailed compilations will be performed, and results with much higher statistical weights than ever will be published. It is certain that the basic statements about the geometry of the TO_4 tetrahedra will not change.

In real structures there is the possibility of systematic distortions of the TO_4 tetrahedra. The influence of the topology and of chemical composition on the tetrahedral shape in the cases of SOD and LTA will be discussed in Sect. 6.

When a single aluminum (Al) atom replaces silicon (Si) in a given aluminosilicate framework, its larger size and lower charge induce strain and the framework has to relax around this defect. A knowledge of this phenomenon is of great importance for the understanding of catalytic properties of zeolites. This point will be discussed in Sect. 8.

In contrast to the T–O bond lengths and OTO bond angles, the TOT angles scatter over a large range of values. This is demonstrated in the histogram shown in Fig. 1 [16], and substantiated by Fig. 2 [17, 18]. In the deformation map in Fig. 2, the potential energy of the Si_2O_7 group in a $H_6Si_2O_7$ molecule is shown,

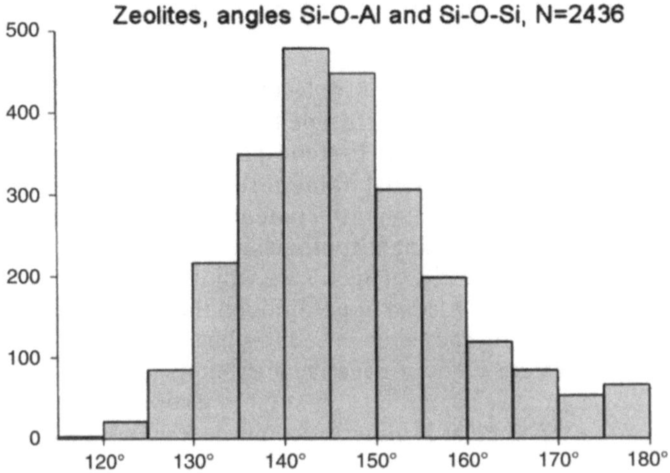

Fig. 1. A histogram of 2436 Si–O–Si and Si–O–Al angles from 317 zeolite crystal structures (from [16])

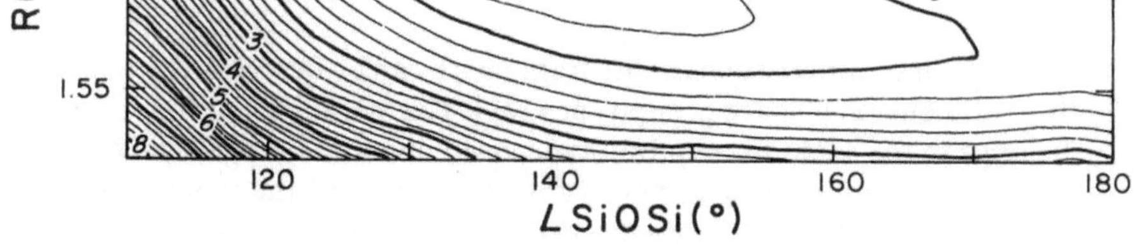

Fig. 2. The potential energy surface for $H_6Si_2O_7$, plotted as a function of the Si–O bond length R(SiO), and the Si–O–Si angle. The energy increments of 0.001 a.u. relative to the minimum (+) are represented by contours. The experimental data points refer to several silica polymorphs (from [39])

as obtained by quantum mechanical methods. It is plotted as a function of two variables, viz. the Si–O bond length and the Si–O–Si angle. From the shape of the valley in the potential energy hyper-surface it is evident that changes in the TOT angle cost only little energy, whereas the narrow cross section of the valley and the steep slopes in the direction of increasing or decreasing Si–O bond length illustrate its low variability. The potential also shows that below about 120° the TOT angles become highly unfavorable, and the structure therefore unstable. The distribution of experimental data points plotted in the theoretical potential hyper-surface suggests that the results obtained for the $H_6Si_2O_7$ molecule can be used to explain the major features of the SiO_4 tetrahedral frameworks as well. However, specific constraints, imposed by the topology of the framework or by the non-framework atoms, have to be taken into account. This can be illustrated by the following example.

Figure 2 suggests that distorting the molecule to a TOT angle of 180° does not cost much energy. If the structure were able to gain enough energy from other sources, then straight angles might become possible. However, the fitting of Si_2O_7 groups into a three-dimensional framework of all-corner-connected tetrahedra will certainly alter the details of the potential energy. The increase in the potential energy when the TOT angle approaches 180° can be much steeper in a framework than in the molecule. In Sect. 7 we will discuss the case of sodalite, where topological constraints set an upper limit on the TOT angle. The question of whether straight TOT bridges exist in zeolites has been a matter of debate in the past. It seems to have found a negative answer, as recent structure refinements have indicated lower symmetry in some questionable cases, where the assumption of the topological symmetry would have required straight angles [19–21].

Many features of the behavior of a tetrahedral framework can be rationalized by a simple rule, viz. that the relative force constants for the T–O bond stretching, the OTO and the TOT angle bending, are in a ratio of 1.0:0.1:0.01 [22]. For many purposes the behavior of a tetrahedral framework can conveniently be represented by a mechanical analogue. It consists of rigid tetrahedra, which are connected by angle brackets. The brackets have a preferred angle of about 140°, but can be strained considerably in response to requirements of the structure [1].

5 Symmetry

The topological symmetry of a zeolite framework can be described by one of the 230 three-dimensional space group types [23]. The symmetry of the real structure belongs to the same space group type, or to one of its subgroups. A knowledge of the true symmetry is necessary for understanding the structural distortions in a zeolite. In some cases the occurrence of commensurate or incommensurate modulated structures requires the use of superspace groups [24]. Irreducible representations of space groups and point groups, as well as isotropy subgroups, are other useful concepts for the discussion of structural distortions [25–27].

Many zeolite framework types possess high topological symmetry. Of the 108 zeolite structure types listed in the atlas [9], 58 have high topological symmetry; they belong to the cubic (12 types), hexagonal (23), trigonal (6), or tetragonal (17) system. Low symmetry space groups occur only in 50 cases (orthorhombic: 35, monoclinic: 15). Zeolites with triclinic topological symmetry do not seem to have been reported yet.

The real symmetry of zeolites is lower than the topological symmetry. The symmetry assignment in the first publication corresponded in only 49 cases to the topological symmetry. This was often revised in later publications; the number of zeolites crystallizing in their topological symmetry is therefore even lower.

The reduction from the topological symmetry to the real space group can often be broken down into smaller steps of group–maximal subgroup relationships. Such a scheme is useful for detecting and discussing relationships between crystal structures, or for the understanding of phase transitions [28]. Phase transitions are common causes for structural distortions in zeolite frameworks. The following is an incomplete list of zeolite frameworks which are known to undergo structural phase transitions without change of the topology [15, 29]: ABW, AEL, AFI, ANA, APD, ATV, EDI, GIS, LTA, MFI, NAT, RHO, SOD. A complete compilation of structural phase transitions in zeolites is still to be published. The powerful method of induced representations of space groups could be applied to the analysis of the atomic distortions of a zeolite-type framework during a phase transition [30, 31].

6
Systematic Tetrahedron Distortion

Usually the TO_4 tetrahedra in zeolite frameworks deviate more or less from the ideal geometry. Various methods have been proposed to describe the distortion. Some methods use basically a scalar quantity to describe the total extent of distortion by summing up, in one way or another, weighted differences between observed and ideal values for distances or angles [12, 13, 32]. Others take the symmetry aspects of the distortion mode explicitly or implicitly into account [33–38]. The reasons why tetrahedra distort and how the distortion relates to the structure and its properties have been discussed in the literature [14, 33, 34, 39, 40].

A case of a particularly strong, systematic tetrahedron distortion was encountered in aluminate sodalites [41]. The AlO_4 tetrahedra are compressed along the ⟨100⟩ directions. This results in two large and four small OTO angles (α and α'). The large angles exhibit record-high values of more than 120°, whereas the small angles are of the order of 105°. Because of the resulting approximate symmetry of the tetrahedron the effect is called "tetragonal tetrahedron distortion" [33].

This strain is the result of topology-induced stress, and its extent depends on the average framework composition: the higher the Al/Si ratio, the larger the angle α. In the sodalite framework each TO_4 tetrahedron belongs to two 4R and to four 6R. Ring tension is strongest in the small 4R. The structure can gain more energy if the ring tension is reduced in these rings. This is why the large angles α are in the 4R.

The ring tension can be reduced more effectively if longer and more ionic T–O bonds allow an easier opening of the angle α. Therefore, aluminate sodalites can display much stronger tetrahedron distortion than their more covalent aluminosilicate homologues. The latter have to change their framework conformation instead (see Sect. 7). The dependence of the tetrahedron distortion angle α on the Al/Si ratio is almost linear [33].

A small influence of the nature of the guest ions on tetrahedron distortion was found in the case of an aluminogermanate sodalite [42].

NMR techniques allow the determination of the distortion of the AlO_4 tetrahedra on a local length scale. For a series of aluminosilicates, the quadrupole coupling constant, QCC, for ^{27}Al was related to the respective tetrahedron distortion [32]. Two parameters were introduced, namely a longitudinal strain $|\alpha|$, and a shear strain $|\psi|$. The QCC was found to depend strongly on $|\psi|$, but is almost independent of $|\alpha|$. A linear relationship between $|\psi|$ and QCC holds for a whole series of various aluminosilicates, including natrolite. An attempt to assign resonances in a given NMR spectrum to symmetrically independent, and thus individually distorted, AlO_4 tetrahedra was hampered by resolution problems.

The problem with the standard MAS-NMR technique is that it cannot cope with second-order interactions. These result in strong line broadening for the resonances of quadrupolar nuclei, such as ^{27}Al. A new technique, DOR-NMR, has allowed the problem to be overcome and has also allowed the registration of spectra of unprecedented resolution from such nuclei [43].

One of the first examples where the new method was able to demonstrate its superiority was CAW, which was known to contain seven independent, highly distorted, AlO_4 tetrahedra [41]. These features made CAW an ideal case for testing the new method. The test was successful, and the dramatic peak sharpening allowed the easy identification of the expected seven central resonances [44]. The linear relationship between QCC and the tetrahedron distortion described by the authors [32] were confirmed and improved.

NMR spectroscopy is complementary to X-ray or neutron diffraction. The NMR spectra of orthorhombic CAW and CAM on the one hand, and of tetragonal SAW and SAM on the other, were shown to match almost perfectly. This indicates that the corresponding tetrahedra have practically identical distortion, in apparent contradiction to the results of X-ray and neutron data [45, 46]. Refinements from low resolution diffraction data had indicated different degrees of distortion for the AlO_4 tetrahedra in each pair, although the differences were below the significance level [47]. The same technique allowed an independent confirmation of the space group of SAS [48].

Mixed crystal effects could be studied in the system $(Ca_{1-x}Sr_x)_8[Al_{12}O_{24}]$-$(WO_4)_2$. Even for $x = 0.1$, heavily broadened ^{27}Al-MAS-NMR spectra were found. The overall shape of the spectra did not change. This is interpreted to be due to superposition of many differently distorted environments of the ^{27}Al nuclei. Different distortions result from different distributions of non-framework atoms around the AlO_4 tetrahedra.

NMR spectra in the middle range of the composition are much broader, but otherwise similar to those of the non-cubic end members. The X-ray patterns of the same samples appear cubic. This indicates that the crystals are cubic on the characteristic length scale of X-rays (say 1000 Å), but non-cubic on the much shorter 10–20 Å length scale of the NMR technique [47].

Another example of a topology-induced tetrahedron distortion occurs in the framework of LTA. Its topology is more complex than that of sodalite, as it contains 8R, in addition to 6R and 4R, the latter being part of double-4R. An idealized LTA framework of cubic symmetry and with identical T atoms was analyzed geometrically. In the model all TO_4 tetrahedra are symmetrically equivalent under space group Pm-3m. The site symmetry of the T atoms contains a mirror plane parallel to {100}, with two oxygen atoms, O1 and O2, being in the same plane. The remaining two atoms O3 and O3' are mirror-related. Symmetrically independent framework oxygen atoms are also topologically distinct: The O1 atom is shared between two 4R and one 8R, O2 belongs to two 6R and one 8R, and O3 to two 4R and one 6R. Note that O2 is not involved in a 4R, in contrast to O1 and O3.

As in the case of sodalites, ring tension is likely to be highest in the small 4R. The structure will thus gain energy if the angles in these rings become larger. This concerns the angles O1–T–O3, O1–T–O3' and O3–T–O3'. The remaining three angles involve the O2 atom, hence they belong either to a 6R (two) or to a 8R. These angles are less susceptible to ring tension and become smaller, presumably in order to keep the average OTO angle close to the ideal one. The two 6R and one 8R around the T–O2 bond make angles of approximately 120° with each other. Therefore the resulting shape of the distorted TO_4 tetrahedron is close to a flattened trigonal pyramid with the O2 atom at the apex, and O1, O3 and O3' forming the base. The trigonal distortion is thus the result of the unequal distribution of 4R, 6R and 8R around the TO_4 tetrahedron. Because it depends mainly on the framework, it is called the "framework contribution to the tetrahedron distortion", or FCD.

In some cases the FCD is superposed by a second effect, called the "cation contribution to the tetrahedron distortion", or CCD. In the simple model both effects are purely additive. It is known that the 6R and the 8R are the favored loci of exchangeable cations, especially in the dehydrated state. The small 4R normally do not contain cations. If all 6R and 8R are fully occupied by cations, as in the case of a dehydrated alkaline zeolite A, the O2 atoms at the apex of the trigonal pyramid are symmetrically surrounded by cations, and there is no reason why the ideal FCD-related trigonal distortion should change significantly.

A different situation occurs in the case of calcium-exchanged dehydrated zeolite A [49]. The 8R are empty, but both 6R are still occupied by cations. The unsymmetrical coordination by Ca attracts the O2 atom towards the bisector between the Ca in the 6R, thereby opening the angle O1–T–O2. Little has changed, on the other hand, for the remaining oxygens and the corresponding angles will change much less, mainly in response to the need to have average values of about 109.5°. The O1–T–O2 angle is more susceptible to different cations and to the way these are distributed around the T–O2 bond than the other OTO angles. The described relationships find their signature in the mean values of the OTO

Table 1. Averages and standard deviations for the TOT angles in the frameworks of 30 different zeolite A species [34]. 4R,6R, 8R denote the type of ring the respective OTO angle is involved in, FCD and CCD indicate "framework contribution" and "cation contribution to the distortion"

TOT angle	Average (e.s.d.)	Type of ring	Controlled by
O1–T–O3 (2×)	111.5 (0.95)°	4R	FCD
O3–T–O3'	110.3 (1.42)°	4R	FCD
O2–T–O3 (2×)	106.3 (1.74)°	6R	FCD
O1–T–O2	110.5 (4.02)°	8R	FCD + CCD

angles, and even more in their standard deviations, as shown in Table 1. These values were calculated from the refined atomic coordinates of 30 different zeolite A species [34]. Note the difference in the standard deviations. The FCD-controlled O1–T–O3 angle varies much less than the CCD-controlled O1–T–O2 angle. The average value of the latter is higher than would be expected from FCD alone. This is because CCD opens the O1–T–O2 angle. Its purely FCD-controlled value would be comparable to that of the O2–T–O3 angles.

Any change in a zeolite A which affects the cation distribution is likely to change the TO_4-tetrahedron distortion. Such changes can be expected to occur as a result of dehydration, adsorption of molecules, heating or compression of the crystal. In particular, dehydration or rehydration will change the strong mitigating effect of water on the cations' influence on the framework. It will not always be easy to detect the resulting changes in the tetrahedron distortion, because the rather subtle effects are superposed by tilt-like framework distortions.

The study of zeolite A species has also revealed systematic bond length distortions. The scatter of the T–O bond lengths increases almost exponentially with the tilt angle (see Sect. 7). From this relationship it seems unlikely that an LTA framework with a tilt angle above 35° will be stable [34].

We cannot see any reason why systematic tetrahedron distortions should not occur in other zeolite frameworks as well. When the number of independent tetrahedra becomes larger than in zeolite A, or even sodalite, the effects are likely to be smaller and less systematic, and thus more difficult to detect. Systematic studies of the relationships between the topology of a given framework and systematic trends for tetrahedron distortions seem to be lacking for all except the two mentioned zeolite frameworks. This is even more true for the variation of tetrahedron distortion with temperature, pressure or chemical composition.

7
Conformational Changes of Zeolite Frameworks

A tetrahedral framework with a given topology can exist in many different conformational states which are distinguished by the relative orientation of adjacent tetrahedra. Mutual re-orientation of the tetrahedra leads to different conformers. One possible way to distinguish between them is to specify the TOT angle at the oxygen atoms.

In contrast to what is insinuated by Fig. 2, the TOT angles are subject to constraints. The topology of the framework forces the rigid TO_4 tetrahedra to be part of rings. The geometry of the tetrahedra has to fit the conditions set by the ring size, and the TOT angles may systematically deviate from the preferred values. In small rings, in particular, this may result in ring tension, which has to be compensated for.

The tetrahedra in a framework are not free to move independently of each other; instead they have to cooperate. The pattern of such a cooperative movement is called a tilt system. A knowledge of its existence is important for understanding the behavior of a given framework structure. The so-called tilt angle specifies the tilting. A tilt angle of 0° refers to the ideal, i.e., the undistorted state. In this zero-tilt system the tetrahedra of the framework have an orientation which has to be defined for each type of framework. This state is often of higher symmetry than the tilted one and may be identified with the ideal structure. The concept of a tilt system helps to understand intuitively the behavior of a framework in terms of a mechanical analogue. This can be useful when the framework expands upon heating, is compressed by high pressure, or is influenced in one way or another by adsorption of guest molecules. In applying the mechanical model one should be aware of its limits. In particular, the assumption of rigid tetrahedra must be reconsidered in each case.

Tilt systems have been studied systematically and in much detail for the octahedral frameworks of the perovskite family [1]. For tetrahedral frameworks a similar systematic approach does not seem to exist. Nevertheless, the very existence of the effect has been known since at least 1930, when Pauling described the "partial collapse" of natrolite and sodalite [50, 51]. The term "collapse" indicates that the volume is involved. Many tilt systems reduce the volume of the unit cell when the tilt angle increases. Such frameworks have been called "collapsible" [52]. However, not all cooperative rotations of the tetrahedra in a framework reduce the unit cell volume. Frameworks exist in which important changes in individual TOT angles do not imply significant volume reduction, because the contributions of individual tetrahedra cancel each other out. While still being "flexible", such frameworks are called "non-collapsible" [16, 52–54]. The difference between a collapsible and a non-collapsible, but flexible, framework is illustrated by the different thermal expansions of zeolite RHO and zeolite A, see Sect. 9.

How far can a framework collapse? Baur [53] proposed three processes that would stop the collapse of a framework: (1) this happens when framework oxygens come too closely into contact with guest species, such as cations, anions, water, or other molecules; the non-framework species act as a kind of spacer for the framework, (2) if the framework is empty, or the cations are very small, the framework cannot continue to collapse when too small TOT angles bring non-bonded oxygen atoms of adjacent tetrahedra into their van der Waals distances where the steep increase of the repulsion term prevents any further approach, and (3) the collapse becomes topologically impossible if the TOT angles around the TO_4 tetrahedra rotate in the opposite sense [52]. The latter is the case of non-collapsible, but flexible, frameworks such as LTA, FAU, or the feldspar framework.

The classical tilt systems of NAT and SOD as described in the literature [50,51] extend to infinity. Recently, other tilt systems have been found which are restricted to a small number of tetrahedra. These local cooperative distortions are ordered in the framework. One example is the so-called shearing in sodalite frameworks [41, 55]. Random local framework distortions of a zeolite framework can occur around single defects, such as a big cation or molecule.

Now we will discuss a few typical examples of tilt systems which occur in zeolite frameworks. NAT belongs to the class of fibrous zeolites in which so-called 4–1 secondary building units form chains along the c-axis. The chains are laterally connected by common oxygen atoms. In a projection down the chains the structure resembles a checkerboard pattern where the black fields represent the cross sections of the chains, and the white fields the tunnels in between. The tunnels accommodate the cations and water molecules. Neighboring chains can rotate in an opposite sense, with the connecting oxygens serving as hinges. The c-axis runs parallel to the chains and is therefore virtually unaffected by the tilt. The perpendicular a- and b-axes are shortened. The tilt also reduces the volume. This model explains the behavior of Na-NAT and, with some restrictions, also that of the K-exchanged form. It fails, however, to explain the measured shortening of the c-axis in the Li-exchanged form. A structure refinement revealed that internal crumpling of the chains is responsible for the shortening. The crumpling along the chains is not cooperative in the sense of a tilt system.

Why does Li-NAT need two distortion mechanisms, while Na-NAT needs only one? The probable explanation is related to the fact that the TOT angles at the hinges are already at their lower limit in Na-NAT. The small Li cations require further reduction in the volume of the unit cell. Crumpling of the chains affects those TOT angles which have not already been involved in the tilt system [56].

The case of NH_4-exchanged NAT demonstrates that weak chemical bonds can also alter drastically the conformation of a framework. The chains are internally twisted and crumpled by hydrogen bonds, but the cooperative tilt system, as in the case of Na-NAT, does not occur [57].

The classical tilt system in the SOD framework [33, 51, 58, 59] is three- rather than two-dimensional as in the case of NAT, and it is depicted in Fig. 3. If the tilt angle $\varphi = 0°$, and only one type of T atom is present, the space group is Im-3m. When φ deviates from 0°, the inversion center is broken, but the cubic symmetry retained, and the space group becomes I-43m. Various symmetry aspects of sodalites have been discussed [60].

The TO_4 tetrahedra in the sodalite framework are systematically distorted, as discussed in Sect. 6. Well-defined relationships exist between the tilt angle, the TOT angle and the tetrahedron distortion angle OTO [33, 58]. Since the tetrahedron distortion angle depends on the average size and electronegativity of the T atoms, and, in addition, the three angles are not independent of each other, there is only a limited range of accessible values for these angles. This is probably the reason why it is difficult for an aluminosilicate sodalite to assume a tilt angle of 0°, i. e., the fully expanded state [33].

Up to now the particular case of shearing has only been found in aluminate sodalites [41, 46], but it can also be expected to occur in the sulfate-bearing aluminosilicate sodalite, hauyne. Its accomplishment is briefly described in Sect. 10 and illustrated in a publication [55]. Shearing is characterized by the following facts: (1) it breaks the cubic symmetry, (2) it breaks translational

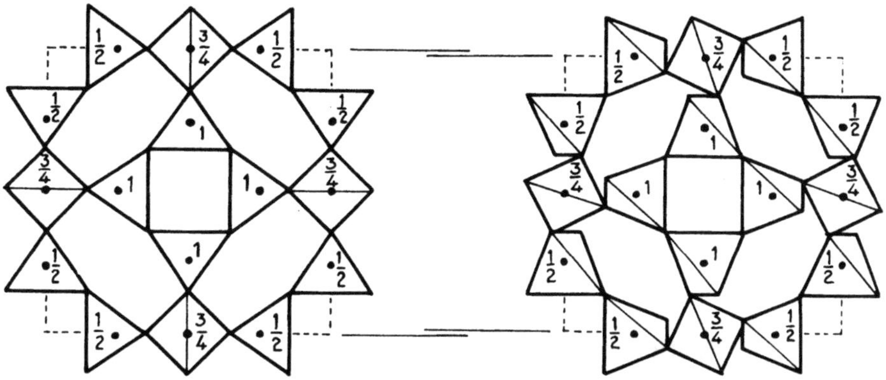

Fig. 3. Schematic drawing of the upper half of the unit cell of sodalite in the fully expanded (*left*) and tilted state (*right*). *Numbers* are relative heights. The partial collapse of the unit cell upon tilting is indicated by the mismatch of the respective lattice parameters in the middle of the drawing (from [59])

symmetry, such that superstructures are formed (see Sect. 10), (3) its occurrence depends on the arrangement of the tetrahedral oxyanions in the sodalite cages, (4) the latter depends on the nature of the cage cations, therefore different shearing patterns occur for different compositions, and (5) it is highly probable that shearing occurs locally and temporally in the cubic phases of aluminate sodalites.

The adaptation of the non-collapsible LTA framework to different chemical compositions cannot be accomplished by changes in the relative orientation of rigid tetrahedra alone, but requires additional tetrahedron distortion. This was shown in a geometrical analysis of the LTA framework [34]. The deviations from the aristotype, with ideal TO_4 tetrahedra being parallel to the cubic unit cell edges, were decomposed into three contributions, viz. (1) bond length distortion of the tetrahedra, (2) distortions of the OTO angles of the tetrahedra, as discussed in Sect. 6, and (3) a tilt system about axes parallel to ⟨100⟩. It was shown that for a given tetrahedron distortion the cubic lattice parameter and the three independent TOT angles depend on the tilt angle. Most importantly, two adjacent TOT angles in the 8R of the framework were shown to change in the opposite sense, in accordance with other work [52, 53]. A scatter plot of the experimental data of 30 chemically different species [34] demonstrates that most of the different zeolite A species are tilted and distorted to such a degree that the lattice parameter becomes maximum, and the TOT angles are almost equal. The distribution of the data points suggests that during chemical reactions, e.g., dehydration or upon heating or compression, the TOT angles have to change quite significantly. In situ structure determinations would help to clarify these points, but are still missing. Increasing the tilt angle away from the relaxed state results in increasing bond length distortion and probably in destabilization of the corresponding zeolite.

Geometrical modeling of some frameworks containing the sodalite cages as building units, viz. SOD, LTA and FAU [61], referred to the description of a cooperative tilt of the tetrahedra in LTA [60]. The model explains quantitatively the major features of the LTA framework.

The analysis of the distortion of the framework of dodecasil 3C-tetrahydrofuran, $[Si_{68}O_{136}] \cdot 4(CH_2)_4O$, exhibited new features. The distortion could be broken down into a tetragonal tetrahedron distortion of one of the SiO_4 tetrahedra, and a localized tilt system. Both distortion mechanisms have their proper mode of symmetry-breaking and their activation is related to phase transitions in this system [62, 63].

8 Structural Relaxation Around Isolated AlO_4 Tetrahedra in a Silica Matrix

Small amounts of Al atoms replacing Si in the TO_4 tetrahedra in highly siliceous zeolites are necessary for their catalytic properties. The substitution produces stress around the AlO_4 tetrahedron. It is interesting to discover how the framework relaxes around such a disturbance. Experimentally, this kind of information is not easy to obtain. When quantum mechanical computational methods were used to determine the re-orientation potentials of the tetramethylammonium cation as guest molecule in the β-cages of tetramethylammonium sodalite, $N(CH_3)_4[AlSi_5O_{12}]$ [64], the desired information was obtained as a by-product.

The periodic model of space group P1 contains three symmetrically independent TO_4 tetrahedra: Si_0O_4 has no direct connection with AlO_4; it is considered to represent the undisturbed silica matrix. Si_IO_4, as intermediate, links Si_0O_4 on all four corners to AlO_4 tetrahedra as second-nearest neighbors. Similar to Si_0O_4, the AlO_4 share corners with Si_IO_4 tetrahedra only. Si_IO_4 has one Si_0O_4, one AlO_4 and two Si_IO_4 as nearest neighbors.

Tetrahedron distortion and local twist are both involved in the reduction of stress. The local twist concerns the AlO_4 tetrahedron, whereas the Si_IO_4 tetrahedra suffer from heavy bond length and angular distortion. The AlO_4 and Si_IO_4 tetrahedra show only topology-induced tetragonal tetrahedron distortion. The twist relaxes back to zero over only three TO_4 tetrahedra, or half the edge length of the cubic unit cell.

The example demonstrates that the sodalite framework can relax over relatively short distances. Bond length distortion, angular distortion, and local twisting contribute to the relaxation. The results are believed to also be significant for the random distribution of the few Al centers in a highly siliceous zeolite framework as long as the distance over which the framework relaxes is shorter than the average distance between the centers. The applicability of the results to other zeolites has still to be tested.

9 Thermal Expansion

The classical model of anharmonic atomic pair potentials is normally insufficient if the thermal expansion of tetrahedral framework structures is to be described. The distortion mechanisms on the microscopic level must be taken into account, as these may contribute to the expansion. They may also cancel out or even result in zero or negative expansion coefficients. Different extremes can be illustrated by the collapsible dehydrated Cs-exchanged zeolite RHO and the non-collapsible dehydrated Na-A zeolite [52]. For the zeolite RHO the cubic lattice parameter varies by 1.84% between 11 and 573 K [65]. For zeolite A in a comparable temperature range between 4.5 and 603 K the change is only 0.03%. Since the measured change between 4.5 and 295 K is 0.22%, the thermal expansion of zeolite A must therefore be negative at higher temperatures [66].

For the thermal expansion of framework structures additional non-linear behavior can be expected. The unfolding of a partially collapsed framework will contribute strongly to expansion. This will continue until it comes to a stop, for instance, when the fully expanded state is attained, or if other forces resist further unfolding. At this point a discontinuity will be expected. If the tetrahedra in a flexible framework become dynamically disordered, vibrational movements may result in apparent bond shortening and negative expansion for lattice parameters and volume.

Many of the concepts underlying these phenomena have been discussed [1]. The effects of not only temperature, but also of pressure or chemical composition, on crystal structures and their microscopic and macroscopic variations are also discussed in detail [67]. An excellent textbook compiling the theories of thermal expansion, experimental techniques and methods for data evaluation has recently been published [68]. The techniques of X-ray diffraction at high temperatures have been discussed [69].

The most extensively studied zeolite family, in regard to thermal expansion behavior, is that of the sodalites. The thermal expansion of 15 different cubic alkali halide aluminosilicate sodalites was determined by precise X-ray powder diffraction experiments [70]. In a few cases of sodalites with big cage anions, strong thermal expansion (about 22×10^{-6} K^{-1}) at lower temperatures changed discontinuously to a regime of low thermal expansion (about 8×10^{-6} K^{-1}) at higher temperatures. There is no doubt that at lower temperatures unfolding of the framework is responsible for the strong expansion, whereas bond anharmonicity contributes practically alone at higher temperatures. However, the nature of the discontinuity in the slope of the thermal expansion curves has been a matter of debate for several years.

One point of view taken was that the framework attains the fully expanded state at the discontinuity [70]. Another idea was that by expansion of the bonds between cage anions and cage cations the latter are pressed against the framework, which expands passively by giving way to this action. The process was believed to come to an end when the cations reach the 0.25, 0.25, 0.25 position in the cubic sodalite structure. This was believed to be a particularly stable one for the cation [58, 71]. This point of view was in turn in conflict with a model of the sodalite structure which showed that the 0.25, 0.25, 0.25 position of the cations cannot be particularly stable. It was proposed instead that the topology of the sodalite framework creates a barrier against further unfolding of the framework. The barrier was believed to be at TOT angles of about 160° [33].

A recent study employing temperature-dependent neutron powder diffraction and Rietveld refinement revealed the concerted movement of the atoms during the unfolding of the framework of the cubic aluminate sodalite $(Ca_{0.5}Sr_{0.5})_8[Al_{12}O_{24}](WO_4)_2$ [47]. The cations were observed passing beyond the 0.25, 0.25, 0.25 position without stopping, and no discontinuity in the thermal expansion curve was observed. The TOT angle remained below 160°. This observation is thus in accordance with that already reported [33].

The effects of ferroelastic phase transitions have been measured for several aluminate sodalites by temperature-dependent determination of the lattice parameters [55]. In SAM, the a = b lattice parameters continue their almost linear downward slope across the cubic → tetragonal transition, apparently without any excess contribution, whereas c decreases strongly below T_c. It follows a power law behavior, as does the volume. This peculiar behavior could be explained by assuming the superposition of two main effects, viz. a spontaneous strain and a volume strain. Both strains cancel out their effects on the a-lattice parameter, but combine for c, thus yielding the observed behavior [72].

A surprising behavior was found for both Sr- and NH_4-exchanged zeolite RHO. The volume of the respective high-temperature forms is about 7% smaller than that of the low-temperature modification. Both modifications are cubic, but differ in the distribution of the large cations. At lower temperatures they are situated at the centers of single 8R. Here, they lock the collapse of the framework which therefore retains a large volume. At a high-temperature phase transition the cations are set free and move to the centers of the double-8R. They no longer block the cooperative anti-rotation of the tetrahedra, the framework is able to collapse and the volume becomes smaller. It is an interesting detail that, in spite of being of the reconstructive type, the transition between the respective high- and low-temperature forms is fully reversible [73, 74].

10 Structural Modulations

In the past two to three decades it has been found that many inorganic or organic materials form periodically distorted, i.e., modulated, structures under defined thermodynamic conditions. Sometimes the wave describing the distortion does not match the underlying lattice periodicity. The modulation is said to be incommensurate. Although perfectly ordered, such a structure no longer has the characteristics and properties of a three-dimensionally periodic crystal. If the ratio between the wavelength of the modulation and the underlying lattice parameters of the undistorted structure (the basic structure) is integral, it is called commensurate. The typical wavelength of a modulation wave is of the order of 50 – 100 Å and, if incommensurate, it depends on parameters such as temperature, pressure or composition.

In this section we will describe the modulations which occur in one class of microporous materials, viz. sodalites, and will then turn to the question as to whether such modulations can also be expected to occur in large-pore zeolites.

All reflections of the reciprocal lattice of a classical crystal can be indexed with three integers h, k, l. For an incommensurately modulated crystal this is not possible, because so-called satellite reflections in the diffraction pattern require an extra contribution to the reciprocal lattice vector $\underline{R}^*$, which can be written (in the simplest case):

$$\underline{R}^* = h\underline{a}^* + k\underline{b}^* + l\underline{c}^* + m\underline{q} \tag{2}$$

with $\underline{a}^*$, $\underline{b}^*$, $\underline{c}^*$ defining the normal reciprocal lattice, h, k, l, m are integers, and

$$\underline{q} = \alpha\underline{a}^* + \beta\underline{b}^* + \gamma\underline{c}^* \tag{3}$$

is the so-called modulation vector. At least one of α, β, or γ must be irrational if the structure can be called incommensurate. The reflections with $m = 0$ are the main reflections. They can be indexed with three integers only, and they form a conspicuous subset of, on average, strong reflections. This is because they represent the Fourier transform of the average structure. Satellites have $m \neq 0$. They carry the information about the structure's deviation from the basic structure, i.e., on the modulation. Their intensity is proportional to the square of the amplitude of the modulation wave. Since the latter is small, satellites usually have weak intensities, $10^{-1}-10^{-4}$ times the intensity of the strongest main reflection, or even lower.

From the occurrence of satellite reflections, it is clear that an incommensurately modulated crystal has lost translational symmetry in at least one direction. In the classical sense it can therefore no longer be called a crystal. On the other hand, the sharp main and satellite reflections in the diffraction pattern reveal long-range order. Because of the loss of the translational symmetry many concepts and tools of conventional crystallography can no longer be applied. This concerns, for example, the lattice, the space group symmetry and the structure factor. The problem can be overcome by embedding the aperiodic crystal into a fictitious higher dimensional space, the so-called superspace. Here, the periodicity is recovered, and the crystallographic concepts can once more be applied after proper adaptation. In the simplest case of a one-dimensional modulation the embedding in a (3+1) dimensional superspace is sufficient to recover the periodicity. The modulated crystal is obtained by calculating the intersection of the superspace with the real space. The theory of symmetry in superspace has been elaborated, and tables of superspace groups and how to determine them from systematic extinctions of main and satellite reflections are available [24]. Programs for structure refinements, calculations of Fourier maps, distance functions, etc., are also available.

In principle, commensurately modulated structures could be treated classically as superstructures in three-dimensional space. In many cases phase transitions between incommensurate and commensurate phases of the same compound exist, the so-called lock-in phase transitions. This demonstrates that commensurate modulations are actually special cases of incommensurate structures, and the superspace approach is more appropriate. Commensurate modulations require special care with respect to the phases of the modulation waves.

The nature of a distortion wave can be quite different. One frequently encountered example is that of substitutional order of atoms, e.g., if atoms, or clusters of atoms, exchange periodically their occupancy on a common site. In extreme cases block-waves can be used to describe such modulations, and the situation can be described as ordering into antiphase domains. Another kind of modulation is found in the so-called displacive distortion waves. Atoms are displaced periodically from their positions in an undistorted structure. The displacement can be perpendicular or parallel to the propagation vector. The corresponding waves are transverse and/or longitudinal. The shape of modulation waves can be simply harmonic or more complicated. In the latter case the shape can conveniently be represented by a Fourier series. The diffraction patterns of substitutional and displacive modulations have been shown to be different [75]. Care must be taken with the interpretation of the diffraction patterns. In studies on modulations occurring in feldspars and sodalites, some of the schematic drawings [75] were misinterpreted and the error propagated through the literature (references on request from the author).

Depending on the nature of the crystal there are various possible physical causes for the occurrence of a modulation. Zeolites are almost exclusively insulators. A model of competing forces as the cause for the occurrence of modulations in insulating crystals has been proposed [76]. For example, a possible competition could occur between ferroelectric polarization and ferroelastic strain. The interactions change sign along the direction of the modulation, and they can be represented as plane waves. The competing interactions are intimately interwoven in such a way that the component waves are in quadrature, i. e., have a phase shift of $\pi/2$. If one interaction has its maximum amplitude, the other one is zero, and vice versa.

In an intuitive one-dimensional model a chain of atoms is placed on an underlying periodic potential. The atoms of the chain are connected by harmonic strings. This model has been treated theoretically in great detail and depth (see, e. g. [77]). For our present purposes the following result is the most important. The ratio between the string constant and the depth of the potential wells determines what kind of modulation results. A commensurate phase will be formed if the string constant is strong relative to the potentials. On the other hand, a chaotic structure will result from random fixing of the atoms in deep potential wells if the string constant is relatively weak. An incommensurate phase is found only if the relative strengths of the sub-systems are well balanced. If the balance changes, e.g., if the string constant becomes smaller, the modulation wave is expected to change as well.

A zeolite-type material where such a kind of competitive interaction has been identified is that of the aluminate sodalite CAW. Even in the earliest publications on CAW and its homologue CAS, the occurrence of superstructure reflections in the X-ray powder patterns was reported [78, 79]. The nature of these extra reflections could not be resolved before crystals of a size suitable for single crystal work became available for CAW. The room-temperature phase was first determined employing a conventional superstructure approach [41], but the interpretation of the highly complex and pseudo-symmetric structure with seven independent AlO_4 tetrahedra remained a puzzle. It could only be understood when it was realized that the structure is in fact commensurately modulated. The modulation could be confirmed by a Rietveld refinement employing the superspace approach [80].

In the conventional description, the space group of CAW is Aba2 and the unit cell is 2(a+b), (−a+b), c, where a, b, c are pseudo-cubic lattice parameters of about 9.3 Å. In the superspace approach the corresponding superspace group is Abm2 (1/2, 0, 0) 0 s 0 and the basic unit cell (a + b), (− a + b), c. The modulation vector is an unusually short one with q = 1/2 a*.

A model for the formation of the structural modulation in CAW is as follows. The structure of aluminate sodalites can be regarded as consisting of three building units: (1) the framework $[Al_{12}O_{24}]^{12-}$, (2) the cage cations (Ca^{2+} or Sr^{2+}) near the center of the 6Rs of the framework, and (3) one tetrahedral oxyanion, e.g. WO_4^{2-}, at the center of the sodalite cage. The symmetry of the tetrahedral cage anions does not fit the latent symmetry of the sodalite cage. Non-spherical cage anions seem to be important for the occurrence of modulations in sodalites. The structural units interact via Coulomb and elastic forces. The positive cage cations are attracted by the negatively charged framework and cage anions, respectively. The interaction between the anions and the framework is repulsive. The requirement of the cage cations to have a sufficiently high number of oxygen atoms in their coordination spheres must also be taken into account.

The resulting structure is depicted in Fig. 4 [81], which actually represents the real structure of the room-temperature phase of CAW [41]. Let us consider the situation in a narrow region where the cations acquire a "good" coordination, e.g., around x = 0.25. This implies that the two neighboring cage anions assume an orientation in which some of their oxygens are at very short distances from the oxygens in the framework. Strong repulsive forces are directed onto the framework. Because these forces act along the diagonal of a 4R of the framework, the latter is locally distorted by shearing (cf. Fig. 3–5 in [55]). Concomitantly, the tilt angle of the affected tetrahedra decreases. This means that the cation acquires favorable coordination only at the expense of a strong distortion of the framework. The situation is reversed in the adjacent region. Opposite orientation of neighboring cage anions now implies a minimum of repulsion with the framework, but an unfavorable coordination number of the cations. The described arrangements of cations and anions repeat periodically, and the framework is periodically distorted as well. Note in Fig. 4 that the two component waves show the required phase shift of $\pi/2$. The local symmetry also changes periodically along the direction of the modulation wave.

We can apply the above-mentioned one-dimensional model to the three-dimensional case of aluminate sodalites. The sodalite framework forms the underlying periodic potential. The atomic chain is replaced by the interacting building units and the string constants by the Coulomb and elastic forces. These forces depend on the nature of the atoms and on their distances from each other. Changes in the composition or in the temperature will therefore change the "string constants". On the other hand, the periodicity of the underlying potential remains fairly constant because it is fixed by the topology of the sodalite framework. Therefore, it is not surprising that all known end members of the aluminate family form modulated structures. The characteristics of the modulations change as a result of different "string constants" and depend on the actual composition and on parameters such as temperature or pressure. For instance, if the Ca in CAW is replaced by Sr the (3 + 3) dimensionally modulated structure

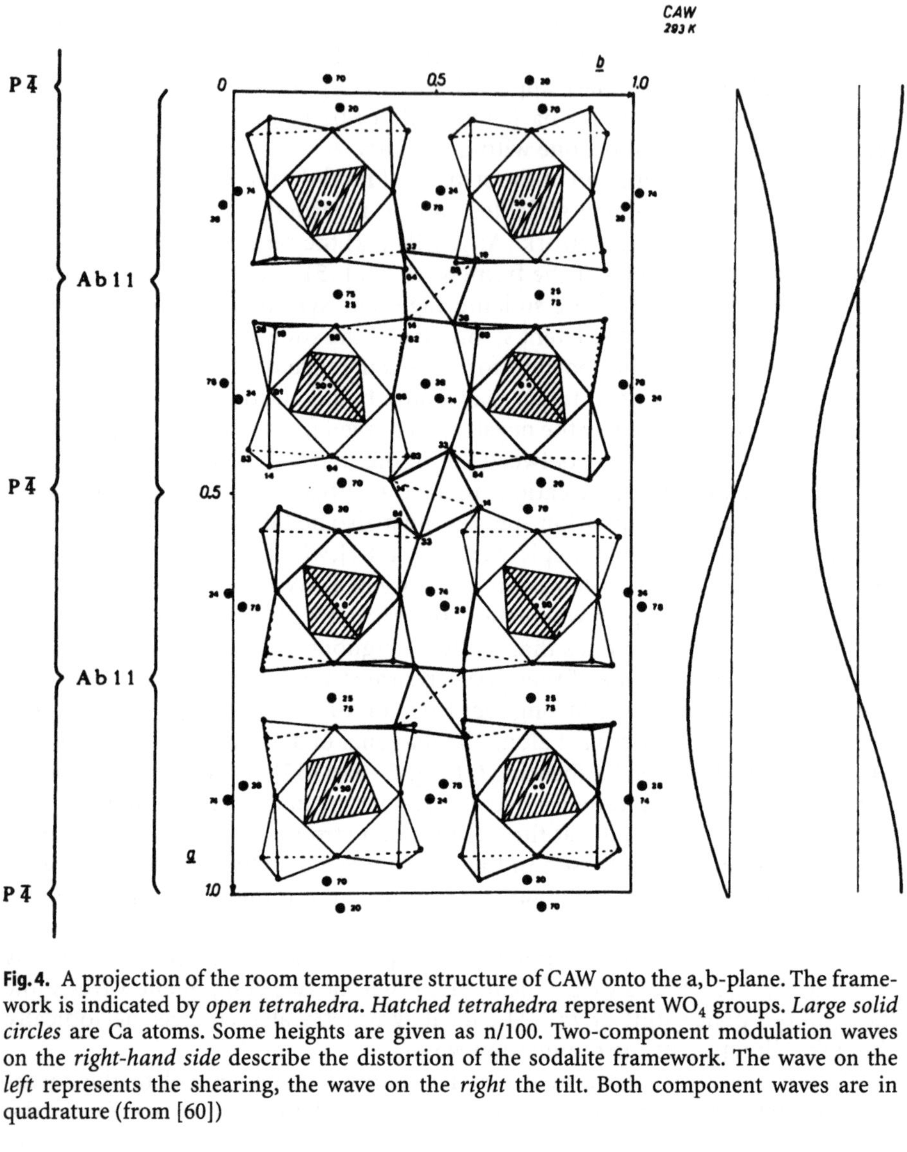

Fig. 4. A projection of the room temperature structure of CAW onto the a, b-plane. The framework is indicated by *open tetrahedra. Hatched tetrahedra* represent WO_4 groups. *Large solid circles* are Ca atoms. Some heights are given as n/100. Two-component modulation waves on the *right-hand side* describe the distortion of the sodalite framework. The wave on the *left* represents the shearing, the wave on the *right* the tilt. Both component waves are in quadrature (from [60])

of SAW is obtained, instead of the (3 + 1) dimensional structure of CAW. This case and others have been discussed in the literature [55, 81, 82].

The subtle balance of forces results in complicated T–X phase diagrams. A recent study on modulations in aluminate sodalites focused on the Ca-rich part of the solid solution system $(Ca_{1-x}Sr_x)_8\,[Al_{12}O_{24}](WO_4)_2$, and employed mainly X-ray, neutron and electron diffraction, as well as solid-state NMR techniques [47, 83]. The resulting phase diagram for this composition range is shown in Fig. 5.

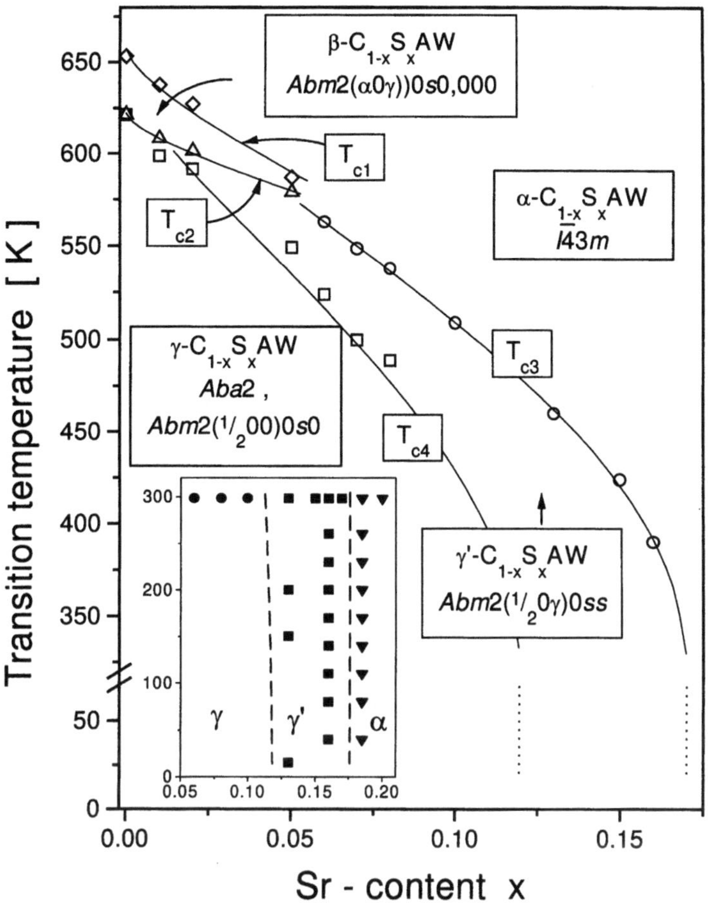

Fig. 5. The X-T phase diagram of the calcium-rich part of the aluminate sodalite solid solution $(Ca_{1-x}Sr_x)_8[Al_{12}O_{24}](WO_4)_2$. The symmetry groups of the phases are given. *Open symbols* represent measured transition temperatures (DSC). The fitted solid curves are discussed in [83]. The low temperature behavior as determined from X-ray experiments is indicated in the inset (from [83])

Below the disordered cubic α-phase all phases are modulated. The room-temperature phase of CAW can incorporate up to 12% of Sr and is denoted γ-$C_{1-x}S_xAW$. The superspace group is Abm2 $(\alpha,0,0)$ 0 s 0. Because $\alpha = 1/2$, this can be reduced to the conventional space group Aba2 and a unit cell with a doubled a-lattice parameter. γ-$C_{1-x}S_xAW$ borders on the fields of β-$C_{1-x}S_xAW$ and γ'-$C_{1-x}S_xAW$. The latter is incommensurate, but otherwise very similar to the γ-phase. The superspace group is Abm2 $(1/2, 0, \gamma)$ 0 s s, with $\gamma \sim 0.05$.

The β-phase of pure CAW has long been known to exist. By electron diffraction it was then established that the modulation is two-dimensionally commensurate. The (3 + 2) dimensional superspace group description is Abm2 $(\alpha, 0, \gamma)$ 0 s 0, 0 0 0, $\alpha = 1/4$, $\gamma = 1/4$. From the different intensity distribution it was deduced that its modulation must differ fundamentally from those of either the γ-or γ'-phase.

The incommensurate γ'-phase does not extend up to pure CAW. This lends support to the speculation that the occurrence of incommensurability in this system requires a contribution from disorder at the cation positions.

The occurrence of satellite reflections and/or diffuse streaks in the diffraction patterns of various aluminosilicate sodalites containing non-spherical cage anions has been known for decades. Examples are hauyne, nosean and lazurite. The phenomena were characterized and interpreted on different levels and with various models [84–94]. A problem with many studies on natural crystals are their chemical and structural inhomogeneities. Well-defined synthetic crystals, such as the aluminate sodalites, are a prerequisite for studies of subtle effects such as structural modulations, which depend so strongly on a delicate balance of forces.

The experience with sodalites can be used in an attempt to appraise the probability of finding modulated phases in large-pore zeolites. The problem translates into the one-dimensional model as having atoms in the chain with much longer distances and, hence, smaller string constants than in sodalites. This means that the solution to the problem will tend to be close to the chaotic regime. Incommensurate or even commensurate modulations are not likely to exist in large-pore zeolites. This estimation agrees with experimental experience. To the best of our knowledge modulated phases similar to those in sodalites have not been reported for large-pore zeolites. The frequent observation, that the cations in large-pore zeolite frameworks occupy various random sites, is also in accordance with the assumption of a chaotic structure.

Attempts were made to find modulated structures in zeolites with slightly larger cages than in sodalites, but still small compared with "real" zeolites. The family chosen was that of the clathrasil dodecasil 3C with MTN topology and neutral guest molecules. The phase transitions were studied in detail and were found to happen at the Γ-point of the Brillouin zone, i.e., no modulation occurred [62, 63]. This fact does not necessarily mean that the possibility of a modulation in such a framework has to be excluded for ever. With stronger interactions, bigger guest species, and purpose-chosen geometry of the guests, it still seems possible to bring about modulations in the MTN framework.

The same is probably true for frameworks with even larger pores. Given that the dimensions of a zeolite framework are practically constant, one has to manipulate the nature of the guest species in order to get the necessary balance of forces for the modulation. There are various conceivable ways of doing so. Using bigger guest molecules or polymerization of guests in channels are just a few.

It seems that an effect of this kind was observed in the case of the adsorption of alkanes in silicalite-1. Hexane and heptane exhibit kinked adsorption isotherms. This is in contrast to short-chain (C_1 to C_5) and long-chain (C_{10}) alkanes which display simple isotherms. Computer simulations suggest that this arises from the interplay between the length of the zig-zag pores and the length of the alkanes. Only when these two are comparable, can the molecules freeze in a configuration that is commensurate with the pore structure [95].

11
Rigid Unit Modes

When the conformation of a tetrahedral framework changes, for example, in response to a change of temperature or pressure, then to a good approximation the TO_4 tetrahedra do not change, and the change of the framework can be described by a "rigid unit mode", or RUM. If we apply the 1.0 : 0.1 : 0.01 rule for the relative force constants of the T–O bonds, the OTO angles and the TOT angles, it is clear that much less energy is needed to distort the framework via changes of the latter than by changes of the former. These low energy modes are thus natural candidates for distortion modes of the framework. The RUM model was first developed for phase transitions in quartz [1, 96].

The idea was further developed by Heine, Dove and coworkers and applied to other frameworks, tetrahedral and others [97–101]. They developed a method to calculate the complete RUM spectrum for any framework silicate. The main trick is that the oxygens which join two rigid TO_4 tetrahedra are treated as split atoms. The two halves are held together by a strong harmonic force, the force constant of which is related to the rigidity of the TO_4 tetrahedra. The RUMs are those modes in which the tetrahedra can vibrate without the split oxygens becoming separated. By using established methods and programs of lattice dynamics, modes can be found in the phonon spectrum which calculate with zero or very low frequency [100]. These are then the RUMs.

In one of their classical papers on RUMs, the authors argue that a framework can only distort if the number of the degrees of freedom exceeds the number of constraints [102]. Since each rigid tetrahedron has six degrees of freedom (three translational, three rotational), and three constraints per oxygen that prevent the split atom from separating, one is left with an equal number of constraints and degrees of freedom. Thus one would expect no deformation mode at all, at least with rigid tetrahedra, and yet the frameworks distort! This is due to the fact that not all constraints are independent, because of symmetry.

The authors have exploited their model very successfully to find answers to a number of puzzling questions, mostly related to structural phase transitions and their characteristics. They have also tackled such problems as to explain why some frameworks have small or even negative thermal expansion, or why some zeolites are collapsible, while others are not. They could also answer the question as to how some frameworks fold locally around cations or molecules without distorting the rest of the structure. They found that this is accomplished by linear combinations of static RUMs. For example in the idealized high-symmetry sodalite structure there is a whole band of RUMs throughout the Brillouin zone. By forming localized wave packets, rings or clusters of oxygens can be formed around a cation without cost of elastic energy. For zeolite A and faujasites an even greater number of localised RUMs can be expected and may be related to the known binding sites in these zeolites.

12
Conclusions

While being topologically quite stable, the tetrahedral framework structures of zeolites and related compounds undergo a multitude of conformational changes. The reasons are, for example, changes of temperature or pressure, hydration or dehydration, loading with extra-framework molecules or cations. Only the effect of temperature has been discussed in this chapter. The remaining points have been omitted owing to space limitations.

Often the conformational changes, or tilt systems, are cooperative, i.e., their correlation length tends to infinity. Localized tilt systems also exist. Structural stress built into a given framework by random defects, e.g., by the replacement of a SiO_4 tetrahedron by AlO_4, has to be removed. It has been shown that the relaxation involves angular and bond length distortion of the tetrahedra, and is relatively short ranged.

The tetrahedra behave as rigid units over large ranges of the changing parameters. However, systematic and sometimes strong tetrahedron distortions occur as the result of topological stresses. The amplitude of this effect depends significantly on the chemical composition, SiO_4 tetrahedra being "harder' than, for example, AlO_4.

With changing external variables, e.g., temperature, the conformational changes may occur more or less smoothly, or with an anomaly in one of the derivatives of the free energy. Usually, the symmetry is broken at such structural phase transitions. This interesting topic has also had to be omitted due to limited space.

Sometimes the broken symmetry corresponds to some general point within the Brillouin zone and, consequently, an incommensurately modulated crystal is formed. The concept of structural modulations can also be applied to cases of commensurate superstructures. Incommensurate or commensurate modulations can be discussed in terms of weak or strong interactions between competing subsystems. If the interactions are still stronger the same concept explains the formation of chaotic structures. In this chapter it is argued that in large-pore zeolites chaotic behavior probably prevails over commensurate or incommensurate modulations. By way of contrast, the latter cases have been found in certain small-pore frameworks, and they are discussed in some detail.

Utilization of concepts and methods of lattice dynamics has resulted in the successful model of rigid unit modes, or RUMs.

From an experimental point of view the structural distortions discussed here are only minor disturbances of an ideal structure. Because of their small amplitude they are difficult to detect. Powerful radiation sources, high resolution instruments, and sophisticated methods of data evaluation are necessary prerequisites to detect the deviations from the ideal state. Despite the small amplitude of the disturbances, their influence on macroscopic properties and, hence, applied aspects of zeolites should not be underestimated.

References

1. Megaw HD (1973) Crystal structures: a working approach. Saunders, Philadelphia
2. Hu X, Depmeier W (1992) Pitfalls in the X-ray structure determination of pseudo-symmetric sodalites and possibly zeolites. Z Kristallogr 201:99
3. Wells AF (1977) Three-dimensional nets and polyhedra. Wiley, New York
4. Wells AF (1979) Further studies of three-dimensional nets. Polycrystal Book Service, Pittsburgh
5. O'Keeffe M, Hyde BG (1996) Crystal structures I: patterns and symmetry. Mineralogical Society of America
6. Bader M, Klee WE, Thimm G (1997) The 3-regular nets with four and six vertices per unit cell. Z Kristallogr 212:553
7. Klein H-J (1996) Prediction of models for tetrahedral frameworks based on the enumeration of graphs. IUCr XVII, Collected Abstracts, C-551
8. Klein H-J (1996) Systematic generation of models for crystal structures. Proc 10th Int Conf on Mathematical and Computer Modelling and Scientific Computing, Boston, MA, July 1995 (Book of Abstracts, p 10) In: Mathematical modelling and scientific computing, vol 6 (special issue)
9. Meier, WM, Olson DH, Baerlocher Ch (eds) (1996) Atlas of zeolite structure types, 4th edn. International Zeolite Association
10. Baerlocher Ch, Hepp A, Meier WM (1976) DLS-76, a program for the simulation of crystal structures by geometric refinement. Institut für Kristallographie, ETH, Zürich
11. Baur WH (1978) Variation of the mean Si–O bond lengths in silicon-oxygen tetrahedra. Acta Crystallogr B34:1751
12. Hill RJ, Gibbs GV (1979) Variation in d(T-O), d(T...T) and ∠TOT in silica and silicate minerals, phosphates and aluminates. Acta Crystallogr B35:25
13. Griffen DT, Ribbe PH (1979) Distortions in the tetrahedral oxyanions of crystalline substances. N Jahrb Miner Abh 137:54
14. Geisinger KL, Gibbs GV, Navrotsky A (1985) A molecular orbital study of bond length and angle variations in framework structures. Phys Chem Miner 11:266
15. Fischer RX, Baur WH (1994) ZeoBase, a database for the display and evaluation of zeolite crystal structure data. 10th Int Zeolite Conf, Garmisch-Partenkirchen, Abstracts, p 570
16. Baur WH (1995) Framework mechanics: limits to the collapse of tetrahedral frameworks. In: Rozwadowski M (ed) Proceedings of the 2nd Polish-German Zeolite Colloquium, Nicholas Copernicus University Press, Toruñ, pp 171–185
17. Gibbs GV, Meagher EP, Newton MD, Swanson DK (1981) A comparison of experimental and theoretical bond length and angle variations for minerals, inorganic solids and molecules. In: O'Keeffe M, Navrotsky A (eds) Structure and bonding I. Academic Press, New York
18. Gibbs GV (1982) Molecules as models for bonding in silicates. Am Mineral 67:421
19. Baur WH (1980) Straight Si–O–Si bridging bonds do exist in silicates and silicon dioxide polymorphs. Acta Crystallogr B36:2198
20. Alberti A (1986) The absence of T–O–T angles of 180° in zeolites. In: New developments in zeolite science and technology. Kodansha, Tokyo
21. Alberti A, Sabelli C (1987) Statistical and true symmetry of ferrierite: possible absence of straight T–O–T bridging bonds. Z Kristallogr 178:249
22. Taylor D (1983) The structural behavior of tetrahedral framework compounds - a review. part I. Structural behavior. Miner Mag 47:319
23. Hahn Th (ed) (1987) International tables for crystallography, vol. A. Space group symmetry, 2nd revised edn. Reidel, Dordrecht
24. Janssen T, Janner A, Looijenga-Vos A, de Wolff PM (1992) Incommensurate and commensurate modulated structures. In: Wilson AJC (ed) International tables for crystallography, vol C. Mathematical, physical and chemical tables. Kluwer, Dordrecht
25. Kovalev OV (1993) In: Stokes HT, Hatch DM (eds) Representations of the crystallographic space groups. Irreducible representations, induced representations and corepresentations, 2nd edn. Gordon and Breach Science Publishers, London

26. Altmann SL, Herzig P (1994) Point-group theory tables. Oxford Science Publications, Clarendon Press, Oxford
27. Stokes HT, Hatch DM (1988) Isotropy subgroups of the 230 crystallographic space groups. World Scientific, Singapore
28. Bärnighausen H (1980) Group-subgroup relations between space groups: a useful tool in crystal chemitsry. Match 9:1456
29. Fischer RX (personal communication)
30. Kahlenberg V (1998) Symmetry analysis and the atomic distortions of the phase transitions in $CsZnPO_4$. Z Kristallogr 213:13
31. Kahlenberg V (1998) Re-investigation of the phase transition in ABW-type $CsLiSO_4$. J Solid State Chem (in press)
32. Ghose S, Tsang T (1973) Structural dependence of quadrupole coupling constant e^2qQ/h for ^{27}Al and crystal field parameter D for Fe^{3+} in aluminosilicates. Am Mineral 58:748
33. Depmeier W (1984) Tetragonal tetrahedra distortions in cubic sodalite frameworks. Acta Crystallogr B40:185
34. Depmeier W (1985) Tilt and tetrahedron distortions in the zeolite A framework. Acta Crystallogr B41:101
35. Klebe G, Weber F (1994) Description of coordination geometry in tetrahedral metal complexes by symmetry-deformation coordinates. Acta Crystallogr B50:50
36. Murray-Rust P, Bürgi HB, Dunitz JD (1978) Distortions of MX_4 molecules from T_d symmetry I: kernel, co-kernel and averaged configurations. Acta Crystallogr B34:1787
37. Murray-Rust P, Bürgi HB, Dunitz JD (1978) Distortions of MX_4 molecules from T_d symmetry II: analysis of PO_4, SO_4 and $AlCl_4$ species. Acta Crystallogr B34:1793
38. Murray-Rust P, Bürgi HB, Dunitz JD (1978) Description of molecular distortions in terms of symmetry coordinates. Acta Crystallogr A35:703
39. Gibbs GV (1982) Molecules as models for bonding in silicates. Am Mineral 67:421
40. O'Keeffe M, Domengès B, Gibbs GV (1985) Ab initio molecular orbital calculations on phosphates: comparison with silicates. J Phys Chem 89:2304
41. Depmeier W (1984) Aluminate sodalite $Ca_8[Al_{12}O_{24}](WO_4)_2$ at room temperature. Acta Crystallogr C40:226
42. Fleet ME (1989) Structures of sodium alumino-germanate sodalites $Na_8(Al_6Si_6O_{24})A_2$, A = Cl, Br, I. Acta Crystallogr C45:843
43. Samoson A, Lippmaa E, Pines A (1988) High resolution solid state NMR–averaging of second-order effects by means of a double-rotor. Molecular Physics 65-4:1013
44. Engelhardt G, Koller H, Sieger P, Depmeier W, Samoson A (1992) ^{27}Al and ^{23}Na double-rotation NMR of sodalites. Solid State Nuclear Magnetic Resonance 1:127
45. van Smaalen S, Dinnebier R, Katzke H, Depmeier W (1991) Aluminate sodalites: structural characterization of the high-temperature phase transitions in $Ca_8[Al_{12}O_{24}](MoO_4)_2$ aluminate sodalite using X-ray powder diffraction. Acta Crystallogr B47:197
46. Depmeier W, Bührer W (1991) Aluminate sodalites: $Sr_8[Al_{12}O_{24}](MoO_4)_2$ (SAM) at 293, 423, 523, 623 and 723 K and $Sr_8[Al_{12}O_{24}](WO_4)_2$ (SAW) at 293 K. Acta Crystallogr B47: 197
47. Többens DM (1998) Untersuchungen zu Struktur und Phasenumwandlungen von Kristallen der Aluminatsodalithgruppe. Dissertation, Universität Kiel
48. Hu X (1992) Untersuchungen zu Strukturen und strukturellen Phasenübergängen im Aluminatsodalith $Sr_8[Al_{12}O_{24}](SO_4)_2$ und einigen Homologen. Dissertation, Technische Universität Berlin
49. Pluth JJ, Smith JV (1983) Crystal structure of dehydrated Ca-exchanged zeolite A. Absence of near-zero-coordinate Ca^{2+}. Presence of Al complex. J Am Chem Soc 105:1192
50. Pauling L (1930) The structure of some sodium and calcium aluminosilicates. Proc Natl Acad Sci USA 16:453
51. Pauling L (1930) The structure of sodalite and helvite. Z Kristallogr 74:213
52. Baur WH (1992) Self-limiting distortion by antirotating hinges is the principle of flexible but noncollapsible frameworks. J Solid State Chem 97:243

53. Baur WH (1995) Why the open framework of zeolite A does not collapse, while the dense framework of natrolite is collapsible. In: Rozwadowski M (ed) Proceedings of the 2nd Polish-German Zeolite Colloquium, Nicholas Copernicus University Press, Toruñ, pp 171–185
54. Baur WH, Joswig W, Müller G (1996) Mechanics of the feldspar-type framework and the crystal structure of Li-feldspar. J Solid State Chem 121:12
55. Depmeier W (1988) Aluminate sodalites–a family with strained structures and ferroic phase transitions. Phys Chem Miner 15:419
56. Baur WH, Kassner D, Kim C-H, Sieber NHW (1990) Flexibility and distortion of the framework of natrolite: crystal structures of ion-exchanged natrolites. Eur J Mineral 2:761
57. Stuckenschmidt E, Kassner D, Joswig W, Baur WH (1992) Flexibility and distortion of the collapsible framework of NAT topology: the crystal structure of NH_4-exchanged natrolite. Eur J Mineral 4:1229
58. Hassan I, Grundy HD (1984) The crystal structures of sodalite-group of minerals. Acta Crystallogr B40:6
59. Taylor D (1972) The thermal expansion of the framework silicates. Miner Mag 38:593
60. Depmeier W (1992) Remarks on symmetries occurring in the sodalite family. Z Kristallogr 199:75
61. Beagley B, Titiloye JO (1992) Modeling the similarities and differences between the sodalite cages (β-cages) in the generic materials–sodalite, zeolites of type A, and zeolites with faujasite frameworks. Struct Chem 3:429
62. Knorr K, Depmeier W (1997) Room-temperature structure and geometrical analysis of the cubic tetragonal phase transition in dodecasil 3C-THF. Acta Crystallogr B53:18
63. Knorr K, Depmeier W (1998) The cubic/tetragonal phase transition in D3C-THF: an optical and X-ray powder diffraction study. J Solid State Chem 137:87
64. Griewatsch C, Winkler B, White J (1998) Ab initio high pressure study of tetramethylammonium sodalite. In: High performance computing in science and engineering, transactions of the HCRS, Springer, Berlin Heidelberg New York
65. Parise JB, Abrams L, Gier TE, Corbon DR, Jorgensen JD, Prince E (1984) Flexibility of the framework of zeolite RHO: structural variation from 11 to 573 K. A study using neutron powder diffraction data. J Phys Chem 88:2303
66. Bennett JM, Blackwell CS, Cox DE (1983) High-resolution ^{29}Si nuclear magnetic resonance and neutron powder diffraction study of Na-A zeolite. Loewenstein's rule vindicated. J Phys Chem 87:3783
67. Hazen RM, Finger LM (1982) Comparative crystal chemistry – temperature, pressure, composition and the variation of crystal structures. Wiley, Chichester
68. Taylor, RE (ed) (1998) In: Ho CY (ed)Thermal expansion of solids. Cindas data series on material properties, vol I-4, Materials Park, ASM International
69. Chung DDL, De Haven PW, Arnold H, Ghosh D (1993) X-ray diffraction at elevated temperatures – a method for in situ process analysis. VCH, Weinheim
70. Henderson CMB, Taylor D (1978) The thermal expansion of synthetic aluminosilicate sodalites, $M_8(Al_6Si_6O_{24})X_2$. Phys Chem Miner 2:337
71. Dempsey MJ, Taylor D (1980) Distance least-squares modelling of the cubic sodalite structure and of the thermal expansion of $Na_8(Al_6Si_6O_{24})I_2$ Phys Chem Miner 6:197
72. Depmeier W, Melzer R, Hu X (1993) The phase transition in $Sr_8[Al_{12}O_{24}](MoO_4)_2$ aluminate sodalite SAM. Acta Crystallogr B49:483
73. Bieniok A, Baur WH (1991) A large volume contraction accompanies the low- to high-temperature transition of zeolite Sr-RHO. J Solid State Chem 90:173
74. Bieniok A, Baur WH (1993) Phase transition of NH_4-RHO to a high-temperature but low-volume form. Acta Crystallogr B49:817
75. Korekawa M (1967) Theorie der Satellitenreflexe. Habilitationsschrift, Universität München
76. McConnell JDC, Heine V (1984) An aid to the structural analysis of incommensurate phases. Acta Crystallogr A40:473

77. Bak P (1982) Commensurate phases, incommensurate phases and the devil's staircase. Rep Prog Phys 45:587
78. Halstead PE, Moore AE (1962) The composition and crystallography of an anhydrous calcium aluminosulphate occurring in expanding cement. J Appl Chem 12:413
79. Saalfeld H, Depmeier W (1972) Silicon-free compounds with sodalite structure. Kristall Technik 7:229
80. Depmeier W, Yamamoto A (1991) Powder profile refinement of a commensurately modulated aluminate sodalite. Mat Sci Forum 79, 763
81. Depmeier W (1991) Modulated phases in aluminate sodalites. In: Methods of structural analysis of modulated structures and quasicrystals, World Scientific, Singapore, pp 397–407
82. Depmeier W (1992) Phase transitions and modulated structures in aluminate sodalites. J Alloys Compounds 188:21
83. Többens D, Depmeier W (1998) Intermediate phases in the Ca-rich part of the system $(Ca_{1-x}Sr_x)_8 Al_{12}O_{24} \cdot (WO_4)_2$. Z Kristallogr 213:522
84. Saalfeld H (1961) Strukturbesonderheiten des Hauyngitters. Z Kristallogr 115:132
85. Schulz H (1970) Struktur- und Überstrukturuntersuchungen an Nosean-Einkristallen. Z Kristallogr 131:114
86. Hassan I (1996) Aluminate sodalite $Ca_8[Al_{12}O_{24}](CrO_4)_2$ with tetragonal and orthorhombic superstructures. Eur J Mineral 8:477
87. Hassan I, Buseck PR (1989) Incommensurate-modulated structure of nosean, a sodalite-group mineral. Am Mineral 74:394
88. Hassan I, Grundy HD (1989) The structure of nosean, ideally $Na_8Al_6Si_6O_{24}SO_4\ H_2O$. Can Mineral 27:165
89. Hassan I, Buseck PR (1989) Cluster ordering and antiphase domain boundaries in hauyne. Can Mineral 27:173
90. Calos NJ, Kennard CHL, Whittaker AK, Davis RL (1995) Structure of Calcium aluminate sulfate $Ca_4Al_6O_{16}S$. J Solid State Chem 119:1
91. Xu H, Veblen DR (1995) Transmission electron microscopy study of optically anisotropic and isotropic hauyne. Am Mineral 80:87
92. Hassan I, Peterson RC, Grundy HD (1985) The structure of lazurite, ideally $Na_6Ca_2(Al_6Si_6O_{24})S_2$, a member of the sodalite group. Acta Crystallogr C41:827
93. Sapozhnikov AN, Vasil'ev EK, Bayliss P (1992) On indexing X-ray diffraction powder patterns of cubic lazurites with an incommensurate-modulated structure. Powder Diffract 7:134
94. Morimoto N (1978) Determination of the modulated superstructures of minerals. Acta Crystallogr A34, Part S4:15
95. Smit B, Maesen TL (1995) Commensurate "freezing" of alkanes in the channels of a zeolite. Nature 374:42
96. Grimm H, Dorner B (1975) On the mechanism of the $\alpha\beta$ phase transformation of quartz. Phys Chem Solids 36:407
97. Giddy AP, Dove MT, Pawley GS, Heine V (1993) The determination of rigid unit modes as potential soft modes for displacive phase transitions in framework crystal structures. Acta Crystallogr A49:697
98. Hammonds KD, Deng H, Heine V, Dove MT (1997) How floppy modes give rise to adsorption sites in zeolites. Phys Rev Lett 78:3701
99. Hammonds KD, Dove MT, Giddy AP, Heine V, Winkler B (1996) Rigid unit phonon modes and structural phase transitions in framework silicates. Am Mineral 81:1057
100. Hammonds KD, Dove MT, Giddy AP, Heine V (1994) Crush: a Fortran program for the analysis of the rigid-unit mode spectrum of a framework structure. Am Mineral 79:1207
101. Dove MT (1997) Theory of displacive phase transitions in minerals. Am Mineral 82:213
102. Dove MT, Heine V, Hammonds KD (1995) Rigid unit modes in framework silicates. Mineralogical Magazine 59:629

Zeolite Type Frameworks: Connectivities, Configurations and Conformations

W. M. Meier[1] and C. Baerlocher[2]

[1] ETH-Zentrum HUT, CH-8092 Zurich, Switzerland; *e-mail:* meier@kristall.erdw.ethz.ch
[2] Laboratory of Crystallography, ETHZ, CH-8092 Zurich, Switzerland; *e-mail:* Ch.Baerlocher@kristall.erdw.ethz.ch

List of Abbreviations 141

1 Introduction 141

2 General Aspects 144

3 Comparison of Silicate and Phosphate Framework Structures 148

4 Framework Densities 148

5 Attempts to Classify Zeolite Frameworks 150

6 Fault Planes 155

7 Loop Configurations and Coordination Sequences 155

8 Framework Conformation 157

9 Are Zeolite Framework Structures Predictable? 159

References 160

List of Abbreviations

DLS distance least squares
FD framework density
FP fault planes
ST structure type
TD topological density

1 Introduction

Zeolite-type framework structures are normally based on fully cross-linked tetrahedra containing Si, Al, P and occasionally other atoms. These are generally called T-atoms. The bridges are invariably formed by oxygen atoms. In a few

Molecular Sieves, Vol. 2

cases, a small fraction of the T–O–T bridges are missing, and in such instances the term 'interrupted framework' is used. The topology of a particular observed crystalline framework is called a structure type (ST) irrespective of composition, distribution of T-atoms, unit cell dimensions, or symmetry. General terms used when dealing with structure types are the *topology* of a 4-connected network or the *connectivity* of a framework, the latter implying a chart or an enumeration scheme of the T-atoms. It should be noted that the actual geometry including bond distances, valence angles etc. does not matter in describing structure types, topologies or connectivities. The same applies to the term *configuration* which has a similar meaning but refers to a local part of the framework and not to the entire 4-connected assemblage. Examples of frequently encountered configurations include cages, apertures, secondary building units (dealt with in Sect. 5), chains, and loop configurations (see Sect. 7 for details). When considering *conformations*, however, the actual geometry of configurations or of an entire framework structure must be taken into account. Unlike conformations of molecular structures, those of tetrahedral frameworks are completely fixed by T–O bond distances, O–T–O and T–O–T angles [1].

The number of established structure types has increased progressively in the last 4 or 5 decades of highly active research in the field of zeolites. The steady growth of the number of STs is shown in Fig. 1 including both silicate- and phosphate-based materials. STs, which were first observed in phosphate-based materials (AlPOs), are shaded in this diagram. Recent progress has been due to both silicate- and phosphate-based materials and the general progress of elucidation of their structures is reflected in Fig. 1. The number of established STs is now clearly in excess of 100. There is no indication that the progressive increase, which is so evident in Fig. 1, might level off in the foreseeable future. Figure 2 demonstrates that powder X-ray diffraction has played an increasingly important role in solving zeolite structures. (This and the possible use of microcrystals combined with synchrotron radiation has been responsible for much of the progress in the structural field). In the early 1950s, when the pioneers in zeolite research started work on the very basis of zeolite chemistry, only 6 STs were known and it was difficult or nearly impossible to learn much from the meager data available at that time.

The structural information which has become available on zeolite-type materials is quite substantial. Fortunately, a compilation of structural data, the ATLAS OF ZEOLITE STRUCTURE TYPES [2], was started in the early 1970s when the volume of available data could still be managed and assessed with the necessary care. This review draws heavily on the 4th edition of this compilation (published in 1996) as well as subsequent updates [3]. The Atlas has been one of the main projects of the IZA Structure Commission. The principal aims of the project have been

a) to provide a reference work based on structural data which has been subjected to critical evaluation,
b) to promote some desirable uniformity in recording such data,
c) to fill in gaps in available data, and
d) to facilitate studies and analyses of data.

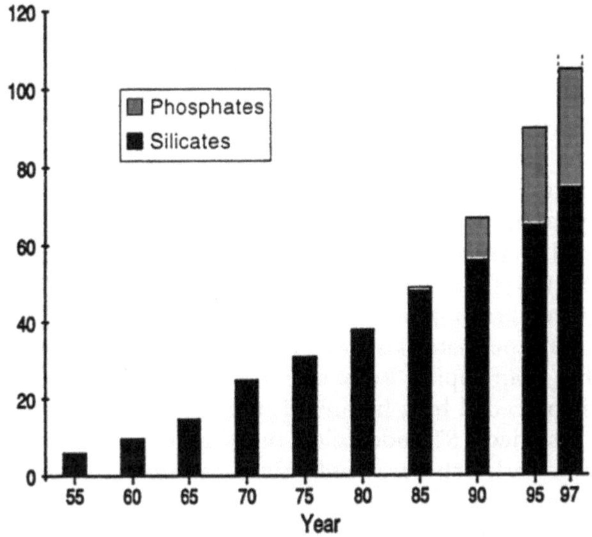

Fig. 1. Development of the number of structure types. The fraction due to phosphates (based on the type material) is indicated in gray

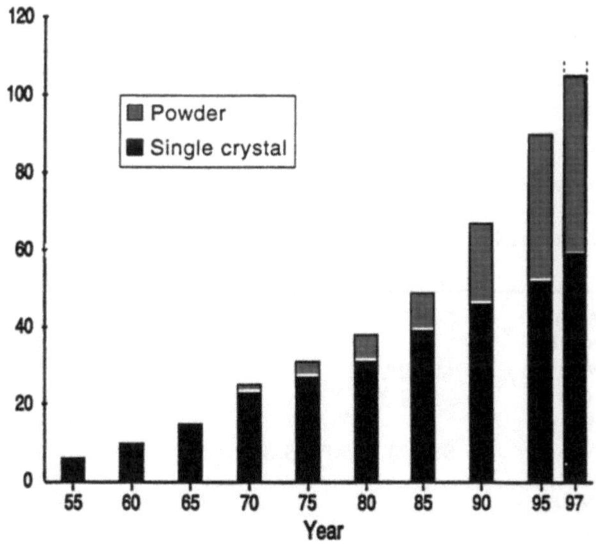

Fig. 2. Development of the number of structure types indicating the fraction established on the basis of XRD powder data (based on [2])

2
General Aspects

All structure types which have been established beyond doubt are assigned a code consisting of three capital letters. This is done by the IZA Structure Commission on behalf of IUPAC. These mnemonic codes should not be confused or equated with actual materials. As indicated before, STs do not depend on composition, distribution of T-atoms (Si, Al, Ga, Ge, B, Be, Zn, Co etc.), cell dimensions or symmetry. They simply denote the topology or the connectivity of a tetrahedral zeolite-type framework which has actually been observed. Hypothetical structures are not included. The codes are generally derived from the names of the type materials (see below) and do not include numbers and characters other than capital Roman letters. For interrupted frameworks the 3-letter code is preceded by a hyphen. However, as a rule, interrupted frameworks are not assigned a ST code unless there is a close resemblance with some zeolite-like framework structures and unless the non-connected T–O bonds are relatively few in number.

Table 1 lists all zeolite structure types approved up to the middle of 1997, their codes, type materials, as well as a selection of isotypes (a more extensive list of isotypes is to be found in the Atlas). The type material is the species first used to establish the ST. It should be noted that this is not necessarily the material which was first synthesized. Isotypes, which are quite frequent, are based on the same ST. Isotypes are even found among zeolite minerals. Numerous examples include analcime (isotypic minerals: leucite, pollucite, wairakite), sodalite (danalite, hauyn, nosean, tugtupite etc.) and stilbite (barrerite, stellerite) although the IMA Commission on New Minerals and Mineral Names has become more restrictive in recent years.

Table 1. Structure types and associated data (based on [2] and [8])

Structure type code	Type material	Selected isotypes [a]	S/P [b]	FD	FD_{Si} [c]	Sieve type [d]
ABW	Li-A(BW)	BePO-ABW, ZnAsO-ABW	S/P	19.0	17.6	*
AEI	AlPO-18		P	14.8	15.1	**
AEL	AlPO-11	MnAPO-11	P	19.1	19.2	*
AET	AlPO-8	MCM-37	P	17.6	18.2	*
AFG	Afghanite		S	15.9	16.9	
AFI	AlPO-5	SSZ-24, CoAPO-5, SAPO-5	S/P	17.5	16.9	*
AFO	AlPO-41		P	19.0	19.2	*
AFR	SAPO-40		P	14.9	15.2	**
AFS	MAPSO-46		P	13.7	14.6	**
AFT	AlPO-52		P	15.2	15.1	***
AFX	SAPO-56	SSZ-16	P	14.7	15.1	***
AFY	CoAPO-50	MgAPO-50	P	12.5	14.1	***
AHT	AlPO-H2		P	18.4	19.2	*
ANA	Analcime	Leucite, Wairakite, AlPO-24	S/P	18.6	19.2	
APC	AlPO-C		P	18.0	17.7	**
APD	AlPO-D		P	19.8	18.0	**

Table 1 (continued)

Structure type code	Type material	Selected isotypes [a]	S/P [b]	FD	FD_{Si} [c]	Sieve type [d]
AST	AlPO-16	Octadecasil	S/P	16.7	15.8	
ATN	MAPO-39		P	18.0	17.8	*
ATO	AlPO-31	SAPO-31	P	19.2	18.8	*
ATS	MAPO-36		P	16.4	16.1	*
ATT	AlPO-12-TAMU	AlPO-33	P	16.7	17.1	**
ATV	AlPO-25	GaPO-ATV	P	20.0	18.9	*
AWW	AlPO-22		P	16.7	16.9	*
*BEA	Beta	Tschernichite, Borosilicate-BEA	S	15.0	15.3	***
BIK	Bikitaite	Cs Aluminosilicate-BIK	S	20.2	18.7	*
BOG	Boggsite		S	15.6	16.1	**
BPH	Beryllophos-phate-H	Linde Q	S/P	16.4	14.6	***
BRE	Brewsterite		S	17.5	18.3	**
CAN	Cancrinite	Tiptopite, basic CAN, ECR-5	S/P	16.7	16.9	*
CAS	Cs Alumino-silicate (A)		S	20.6	18.8	*
CGF	Co-Ga-phos-phate-5		P	17.4	19.0	**
CHA	Chabazite	AlPO-34, MeAPO-47, Phi	S/P	14.6	15.1	***
-CHI	Chiavennite		S	20.9	19.9	*
-CLO	Cloverite		P	11.1	11.1	***
CON	CIT-1	SSZ-26/SSZ-33	S	16.1	15.7	***
CZP	'Chiral' Zn-phosphate		P	16.7	21.3	*
DAC	Dachiardite	Svetlozarite	S	17.3	17.5	**
DDR	Deca-dode-casil-3R	Sigma-1, ZSM-58	S	17.6	17.9	*
DFO	DAF-1		P	14.1	14.9	***
DOH	Dodecasil-1H		S	18.5	17.0	
EAB	TMA-E(AB)	Bellbergite	S	15.4	16.0	**
EDI	Edingtonite	K-F, Linde-F	S	16.6	16.3	***
EMT	EMC-2		S	12.9	13.3	***
EPI	Epistilbite		S	18.0	17.7	**
ERI	Erionite	AlPO-17, LZ220	S/P	15.6	16.1	***
EUO	EU-1	TPZ-3, ZSM-50	S	18.2	17.1	*
FAU	Faujasite	X, Y, SAPO-37, ZnPO-X	S/P	12.7	13.3	***
FER	Ferrierite	Sr-D, FU-9, NU-23, ZSM-35	S	17.7	17.5	**
GIS	Gismondine	Garronite, Na-P, MAPO-43	S/P	15.4	16.4	***
GME	Gmelinite		S	14.6	15.1	***
GOO	Goosecreekite		S	17.6	19.0	***
HEU	Heulandite	Clinoptilolite, LZ219	S	17.0	17.5	**
IFR	ITQ-4		S	17.0	17.2	*
JBW	Na-J(BW)		S	18.6	18.8	*
KFI	ZK-5	P, Q	S	14.7	15.0	***
LAU	Laumontite	Leonhardite, CoGaPO-LAU	S/P	17.7	18.0	*
LEV	Levyne	LZ-132, NU-3,SAPO-35	S/P	15.2	15.9	**
LIO	Liottite		S	15.7	17.6	
LOS	Losod	Bystrite, LiBePO-LOS	S/P	15.8	16.9	
LOV	Lovdarite	synthetic Lovdarite	S	18.4	16.8	***

Table 1 (continued)

Structure type code	Type material	Selected isotypes [a]	S/P [b]	FD	FD_{Si} [c]	Sieve type [d]
LTA	Linde type A	SAPO-42, ZK-4, GaPO-LTA	S/P	12.9	14.2	***
LTL	Linde type L	Perlialite, (K,Ba)-G,L	S	16.4	16.7	*
LTN	Linde type N	NaZ-21	S	15.2	17.0	
MAZ	Mazzite	Omega, ZSM-4	S	16.1	16.7	*
MEI	ZSM-18		S	14.3	14.	***
MEL	ZSM-11	Silicalite-2, Borolite-D, TS-2	S	17.7	17.4	***
MEP	Melanophlogite	synthetic MEP	S	18.9	17.9	
MER	Merlinoite	K-M, Linde W	S	16.0	16.4	***
MFI	ZSM-5	Silicalite-1, Borolite C, TS-1	S	17.9	18.4	***
MFS	ZSM-57		S	18.2	17.4	**
MON	Montesommaite		S	18.1	17.7	**
MOR	Mordenite	Ca-Q, LZ-211, Zeolon	S	17.2	17.0	**
MTN	ZSM-39	Dodecasil 3C, Holdstite	S	18.6	17.2	
MTT	ZSM-23	EU-13, KZ-1	S	20.0	18.2	*
MTW	ZSM-12	CZH-5, VS-12	S	19.4	18.2	*
MWW	MCM-22	ERB-1, ITQ-1, PSH-3, SSZ-25	S	16.8	15.9	**
NAT	Natrolite	Scolecite, Mesolite, Gonnardite	S	17.8	16.2	***
NES	NU-87	Gottardiite	S	17.7	16.4	**
NON	Nonasil	ZSM-51, Borosilicate NON	S	19.3	17.6	
OFF	Offretite	TMA-O, LZ-217	S	15.5	16.1	***
OSI	UiO-6		P	18.8	17.7	*
-PAR	Partheite		S	18.2	19.5	*
PAU	Paulingite	ECR-18	S	15.5	15.9	***
PHI	Phillipsite	Harmotome, Wellsite, ZK-19	S	15.8	16.4	***
RHO	Rho	Pahasapaite, LZ-214	S/P	14.3	14.5	***
-RON	Roggianite		S	15.6	19.1	*
RSN	RUB-17		S	16.8	16.8	***
RTE	RUB-3		S	17.3	17.2	*
RTH	RUB-13		S	16.6	16.1	**
RUT	RUB-10	NU-1, TMA-silicate-RUT	S	18.7	18.1	
SAO	STA-1		P	13.9	14.2	***
SGT	Sigma-2		S	17.8	17.5	
SOD	Sodalite	Tugtupite, basic SOD, AlPO-20	S/P	17.2	16.7	
STI	Stilbite	Barrerite, Stellerite	S	16.9	16.7	*
TER	Terranovaite		S	17.1	17.0	**
THO	Thomsonite		S	17.7	15.7	***
TON	Theta-1	ISI-1, KZ-2, NU-10, ZSM-22	S	19.7	18.1	*
VET	VPI-8		S	19.8	20.2	*
VFI	VPI-5	AlPO-54, MCM-9, H1	P	14.2	14.5	*
VNI	VPI-9		S	16.7	17.6	**
VSV	VPI-7	Gaultite	S	17.1	16.8	***
WEI	Weinebeneite		P	18.1	16.5	**
-WEN	Wenkite		S	17.0	16.6	***
YUG	Yugawaralite	Sr-Q	S	18.3	18.0	**
ZON	ZAPO-M1	GaPO-DAB, UiO-7,	P	17.0	18.0	**

[a] A more extensive list of isotypes is given in the Atlas [2].
[b] Silicate and/or phosphate.
[c] Famework density based on a pure (hypothetical) SiO_2 framework (see in [3]).
[d] 1-, 2-, 3-Dimensional channel system indicated by the number of asterisks.

The characteristic channel systems in zeolite-type structures cannot be adequately described in a simple manner. An unrefined but frequently adopted way is to distinguish 1-, 2- and 3-dimensional channel systems only. Such a classification is included in Table 1 and is only meaningful for rough surveys. For anything more detailed, the Atlas should be consulted. The size and shape of the apertures vary greatly, from an 8-ring opening to 9-, 10-, 12-, 14-, 18-, and 20-ring structures (counting the number of T-atoms in the shortest circuit, which is not trivial in the case of large and highly puckered rings). A survey of the observed apertures is given in Fig. 3. 6-Ring systems are of limited interest in applications involving sorption and are not considered as a rule. For more details on the channels including crystallographic free diameters and the shape of the apertures the reader is again referred to the Atlas.

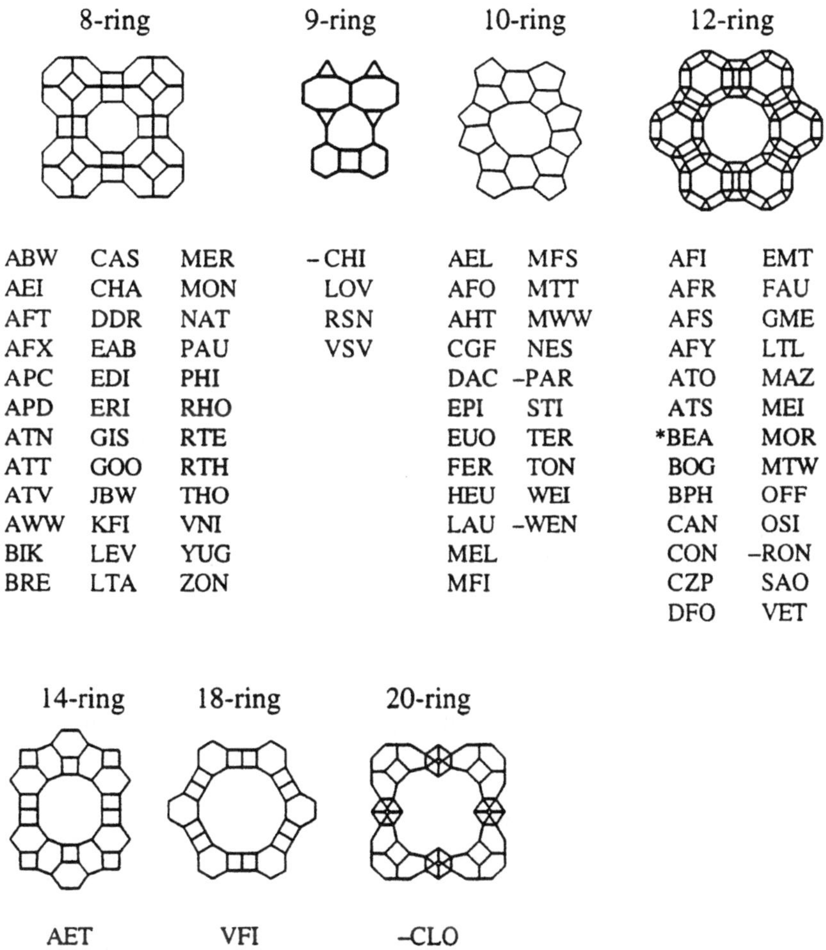

Fig. 3. A listing of the structure types according to their largest aperture. A simplified projection of a typical framework is given to illustrate each aperture

3
Comparison of Silicate and Phosphate Framework Structures

Zeolite structure types include both silicates and phosphates. From a chemical point of view these constitute two distinctive groups of microporous materials. They have been listed using the abbreviations S or P in the respective column of Table 1. The listing shows that there are three rather than two distinct groups to be considered. Apart from those associated with silicates (59%) and phosphates (26%) there is a sizable number of STs which have been found to occur both in silicates and phosphates (15%). Detailed structural investigations of isotypic pairs of this latter group (AFI e.g. [4]) provide a deeper insight into the stereochemistry of these materials and still more is to be expected.

Soon after the first AlPO-structures (including AFI) were solved it was recognized that the T-atoms in these frameworks are ordered, obviously for reasons of local charge balance. As a consequence these phosphate systems can only form frameworks containing even-membered rings which is in line with the results of all structural investigations carried out so far. Thus, 5-membered rings, which are so common in mineral zeolites and purely synthetic silicates (particularly high-silica zeolites), are not expected to form in microporous phosphates. Some further comparisons of zeolitic silicates and phosphates can be found in Sect. 7.

4
Framework Densities (FD)

Zeolites and zeolite-like materials do not comprise an easily definable family of crystalline solids. Thus, in several well-known mineralogical reference works [5] analcime and bikitaite, for example, are not listed under zeolites. A straightforward criterion for distinguishing zeolites and zeolite-like materials from denser tectosilicates is best based on the framework density (FD), the number of T-atoms per 1000 A^3. FD values for the various structure types are listed in Table 1. The FD is obviously related to the pore volume but does not reflect the size of the pore openings. Normally FD-values refer to the type material and depend somewhat on composition since they rest on lattice constants. They can be normalized on the basis of a (theoretical) all-silica framework structure having optimal bond distances and angles [3]. These normalized FD-values are also given in Table 1 for comparison. For most structure types FDs do not differ by more than a few percent. Larger differences like the 10% observed in NAT, for example, are due to lack of rigidity of the framework (see Sect. 8 for more information in this respect).

In Fig. 4 the distribution of the FD-values for porous and dense frameworks whose structures are well established are plotted against the smallest rings present (adapted from [2]). Strictly speaking the boundaries defined in Fig. 4 for the framework densities apply to fully cross-linked frameworks only. A gap (marked by the hatched area in Fig. 4) is clearly recognizable between zeolite-type and dense tetrahedral framework structures. The boundary ranges from just under 20 to over 21 T-atoms per 1000 A^3, depending on the smallest rings present. The

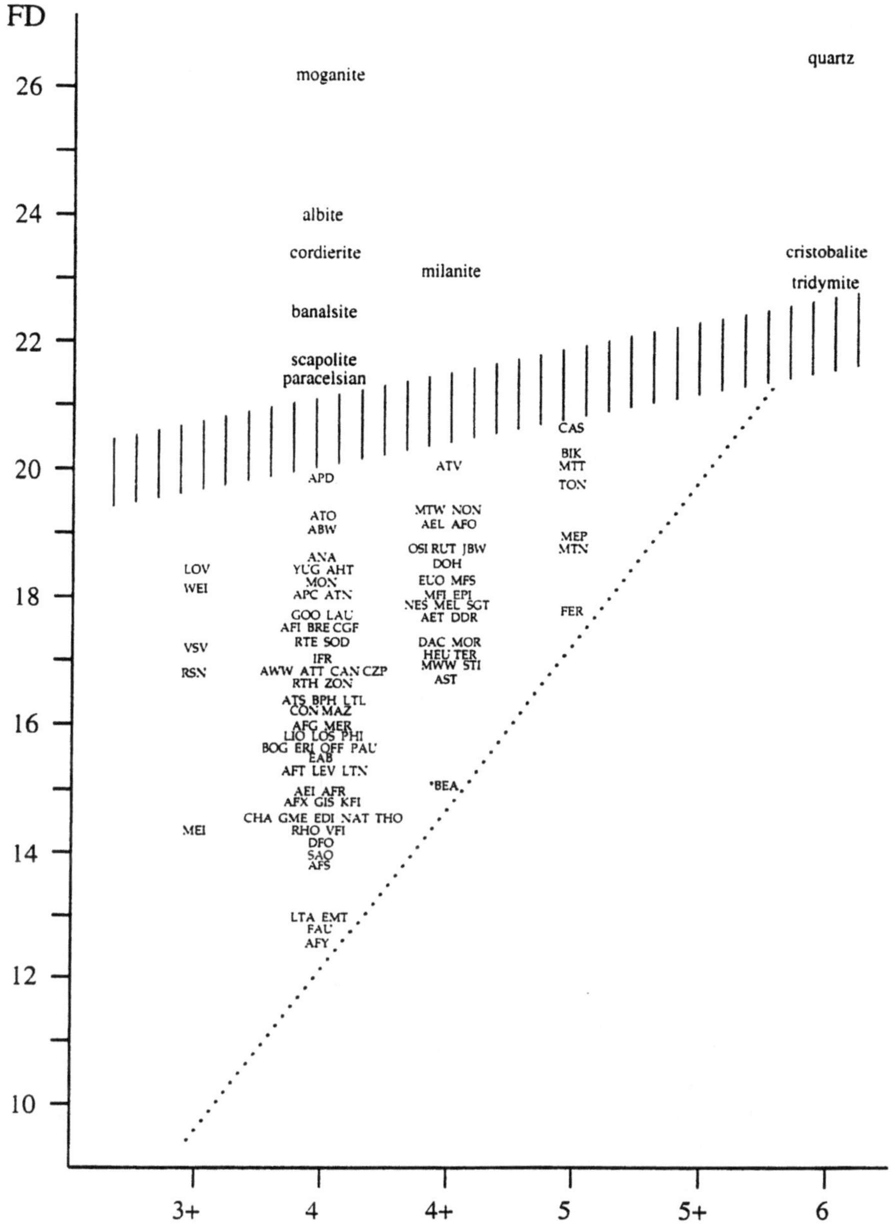

Fig. 4. Plot of FD vs. smallest ring(s)

significance of the smallest rings in this context was first noted by Brunner and Meier [6].

As can be seen in Fig. 4, the largest number of STs are in category '4'. That is, all of the T-atoms in these structure types are associated with 4-rings and the 4-rings are the smallest loops in the structure. In the case of category '4+', the smallest loop is also a 4-ring, but some of the T-atoms are part of larger rings only. This diagram defines a 2-dimensional range for zeolite-type structures. A lower boundary for the FDs is also apparent. One important conclusion derived from this plot has been that low-density frameworks are likely to contain 3-rings. This hypothesis has been tested by hypothetical structures and 3-ring structures having FDs below 10 could in fact be generated [7]. Originally, the only zeolitic material containing 3-rings was lovdarite, a beryllosilicate. New structure types, in particular MEI and some zincosilicates (RSN, VNI, VSV), are very significant for they show that silicates can also form tetrahedral nets containing 3-rings, without Be being needed. This bodes well for the synthesis of low-density materials based on 3-ring structures in the silicate field.

Considering the sharp increase in the number of structure types it is perhaps noteworthy that the boundaries in the FD plot shown in Fig. 4 have not been changed since it was first conceived almost 10 years ago.

5 Attempts to Classify Zeolite Frameworks

As was recognized quite early, a number of frameworks are obviously related and the most important examples of these are to be found in Table 2. This table contains only the STs listed in the 3rd edition of the Atlas [8].

The meaning of the designations used is as follows:

cha frameworks made up of single and/or double 6-rings (occasionally called the "ABC family")
cla clathrasils i.e. cage structures having no openings larger than 6-rings
fau frameworks consisting of sodalite or larger alpha cages
heu structures resembling that of heulandite
mor structures resembling that of mordenite
nat structures formed by chains found in natrolite
phi frameworks formed by chains of 4-rings ("UUDD and UDUD chains" originally conceived by J.V. Smith) [9]

The entries found in Table 2 show quite clearly that these well-known "families" of structures cover only about half of the STs presently known.

Zeolite frameworks can be thought of as consisting of finite or infinite (i.e. chain- or layer-like) component units. The finite units or configurational constituents which have been found to occur in tetrahedral frameworks are shown in Fig. 5 (adapted from [8]). These secondary building units (SBU), as they are normally called, contain up to 16 T-atoms and are derived under the assumption that the entire framework is made up of *one* type of SBU only. It should be noted that these constituent units must be non-chiral to be of use, or else additional

Table 2. Related structure types, classification systems, and fault planes (For explanatory notes see Sects. 5 and 6)

Structure type code	Examples of related ST	SBUs	Breck's classif.[a]	Gottardi's classif.	Fault planes
ABW		4, 6, 8	(B1)		(010), (011)
AEI		4, 6, 6–6	(B4)		(001)
AEL		6–2			(010)
AET		6–2			(100), (010)
AFG	cha	4, 6	(B2)		(001)
AFI		4, 6	(B2)		(210)
AFO		4–1			(010)
AFR		4, 6–2			
AFS		6≡1			(001)
AFT	cha	4, 6,6–6	(B4)		(001)
AFY		4, 4–4	(B3)		
ANA		4, 6, 6–2	B1	G2	(100)
APC	phi	4, 8	(B1)		(010)
APD		4, 8, 6–2	(B1)		(100), (001)
AST		4–1			
ATN		4, 8	(B1)		
ATO		4, 6	(B2 ?)		
ATS		4, 6, 6–2	(B1 ?)		
ATT		4, 6–2	(B1)		(001)
ATV		4	(B1)		(100), (010)
AWW		4, 6	(B1 ?)		
*BEA		two 4 + 5–3			(001)
BIK	mor	5–1	B6	G5	
BOG		4, 6	(B1 ?)		(011)
BPH		6≡1			(001)
BRE	heu	4	B7	G6	(010)
CAN	cha	4, 6	B2		(001)
CAS		5–1	(B5)		
CHA	cha	4, 6, 6–6	B4	G4	(001)
-CHI		5–2			
-CLO		4, 4–4	(B3)		
DAC	mor	5–1	B6	G5	(010)
DDR		5–1			
DOH	cla	5 + two 5–1			
EAB	cha	4, 6	(B2)		(001)
EDI	nat	4=1	B5	G1	(110)
EMT	fau	4, 6, 6–6	(B4)		(001)
EPI	mor	5–1	B6	G5	(010)
ERI	cha	4, 6	B2	G4	(001)
EUO		4 + four 5–1			
FAU	fau	4, 6, 6–6, 6–2	B4	G4	(111)
FER	mor	5–1	B6	G5	(010)
GIS	phi	4, 8	B1	G3	(101), (011)
GME	cha	4, 6, 8, 6–6	B4	G4	(001)
GOO		6–2			
HEU	heu	4–4=1	B7	G6	(010)
JBW		6			
KFI	fau	4,6,8,6–6,6–2	B4		

Table 2 (continued)

Structure type code	Examples of related ST	SBUs	Breck's classif.[a]	Gottardi's classif.	Fault planes
LAU		6, 6–2	B1	G2	
LEV	cha	6	B2	G4	(001)
LIO	cha	6	(B2)		(001)
LOS	cha	6, 6–2	B2		(001)
LOV		4 + spiro-5			(100)
LTA	fau	4, 8, 4–4, 6–2	B3		
LTL		6	B2		
LTN		4, 6, 6–2			
MAZ		4, 5–1	B2	G3	(010)
MEI		3 + two 6≡1			
MEL	mor	5–1	(B6)		(100)
MEP	cla	5 + three 5–1			
MER		4, 8, 8–8		G3	(100)
MFI	mor	5–1	(B6)		(100)
MFS		5–1			
MON		4	(B1)		(001)
MOR		5–1	B6	G5	(010), (100)
MTN	cla	5 + two 5–1			
MTT		5–1	(B6)		
MTW		4 + four 5–1			
NAT	nat	4=1	B6	G1	(100)
NES		5–3			
NON	cla	5 + 6			
OFF	cha	6	B2	G4	(001)
-PAR		4	(B1)		
PAU	fau	4	B1	G3	
PHI	phi	4, 8	B1	G3	(010)
RHO	fau	4,6,8,8–8,6–2	(B1)		
-RON		4	(B1)	G2	
SGT	cla	5–3			
SOD	cha, fau, cla	4, 6, 6–2	B2	G6	(111)
STI	heu	4–4=1	B7	G6	(010)
THO	nat	4=1	B5	G1	(100), (010)
TON		6, 5–1			
VFI		6	(B2)		
-WEN		two 6 + 8			(100), (210)
YUG		4, 8	B1	G2	(010)

[a] Entries in parentheses refer to new structure types which have also been classified based on Breck's criteria. Question marks indicate considerable ambiguities.

specifications are required making the scheme impractical. Furthermore, a unit cell always contains an integral number of SBUs. In some instances combinations of SBUs have been encountered, i.e. the respective framework cannot be generated from *one* type of SBU only. Examples include LOV, MEP and other clathrasil-type frameworks. The SBUs for the various structure types are listed in Table 2.

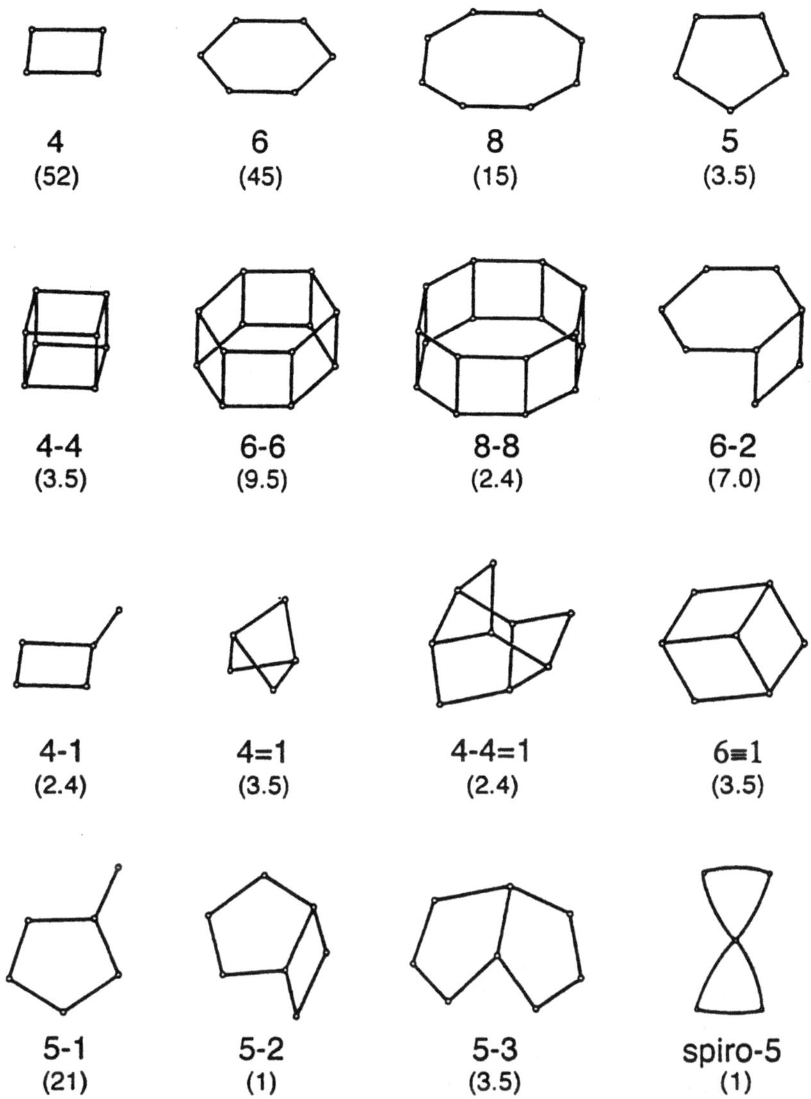

Fig. 5. SBUs (adapted from [8]) including frequency of occurrence in % (in parentheses)

Early classification schemes of zeolite framework structures date back to the 1960s when it did not seem to be a pressing problem, as the number of known STs at that time was only around 25. It is remarkable that even at that time the family of zeolite structure types was divided up into no less than 6 or 7 groups based on SBUs and other structural similarities. This is still evident in Breck's classification [10]. There were "35 known, different framework topologies and an infinite number possible" in 1973 according to Breck. (Based on present knowledge the number stated should have been 30; v.i. Sect. 8). These structure types were classified into 7 groups which are listed in Table 2 using the notation B1 to B7 for each structure type considered. Breck's classification scheme was based on the following set of SBUs:

Group 1	SBU 4	(single 4-ring)	S4R
Group 2	SBU 6	(single 6-ring)	S6R
Group 3	SBU 4–4	(double 4-ring)	D4R
Group 4	SBU 6–6	(double 6-ring)	D6R
Group 5	SBU 4–1		
Group 6	SBU 5–1		
Group 7	Group 4–4–1		

Structure types which were established later have also been classified according to Breck's scheme whenever possible, giving the group name in parentheses. The listing shows that Breck's classification was quite acceptable in the 1970s but in many cases (over 30%) could no longer be applied in 1992, nor could it be readily extended.

A later attempt to classify the structure types by Gottardi and Galli [11] conducted on similar grounds (but including chains and sheets) led to 6 groups but was at least initially limited to natural zeolites. This classification is also listed in Table 2 and labels used are G1 to G6. The groups were defined as follows:

Group 1	natrolite group (4 = 1)
Group 2	singly connected 4-ring chains
Group 3	doubly connected 4-ring chains
Group 4	6-rings (6 and 6–6)
Group 5	mordenite group (hexagonal sheet with "handles")
Group 6	heulandite group (4–4 = 1)

Again this classification can only be applied to a limited number of the STs presently known. There is considerable but not complete overlap with Breck's classification.

The present account covers just two representative examples out of many more attempts to classify the known zeolite framework topologies. From a general point of view an acceptable classification system should (i) group together comparable structure types, (ii) be based on criteria which can be applied unambiguously, and (iii) be applicable to all structure types without having to set up an excessive number of groups, which would of course serve no useful purpose. SBUs alone fail to provide a classification system which would satisfy all the aforementioned criteria. Thus, in view of criterion (i), ANA and PAU should not appear in the same group (B1) simply because they both happen to have an SBU in common, the single 4-ring. As is quite evident from Table 2, in many instances the choice of an SBU is ambiguous. The same applies to other configurations which may seem typical in a particular topology. A different approach to the classification of STs would be based on the use of the smallest ring(s) (see Fig. 4), an interesting feature discussed in the last section. The foregoing criteria (ii) and (iii) would be readily fulfilled, there are no ambiguities involved in assigning an ST to a particular group and the number of groups would be remarkably small (4 at present). However, to have FAU, ERI, NAT etc. in the same group is untenable. A structural classification system which satisfies all three criteria is indeed yet to be found.

6 Fault Planes

Structural faults are quite abundant in zeolite crystals and can in particular cases (e.g. zeolite beta) pose major problems in resolving details of the structure. The existence of fault planes (FP) means that layer-like segments of the framework can be stacked in more than one way. Figure 6 serves in a schematic way as an illustration of a potential FP. Likely FPs have been listed for known structure types in Table 2 (taken from [8]). Well-known examples include the "ABC family" of structures labeled *cha* in Table 2. The AABB stacking of 6-rings in GME is hardly ever observed without some stacking fault (CC-layer), which greatly affects the porosity in this instance. A different example of importance is that of FAU (cubic) and EMT (hexagonal) in which faults do not affect the porosity greatly, if at all. The faults in MFI and MEL are somewhat different in nature but like all the others considered they are conservative, i.e. layers or segments of the structure are just stacked differently without adding or leaving out any part. A possible FP means that the respective framework can easily contain defects of this kind, which may be difficult to detect directly if they are few in number. On the other hand, a high concentration of faults gives rise to structural "intermediates". These include Linde T (ERI-OFF) and FAU-EMT intermediates [12] (e.g. CS1, ECR-30, ZSM-3 and ZSM-20), to quote just two examples. The listing in Table 2 shows that many more framework structures are prone to faulting than generally expected (around 60% of all STs).

Finally, a number of related structures (polytypes) can be readily postulated on the basis of possible faulting (see Sect. 9 for more details).

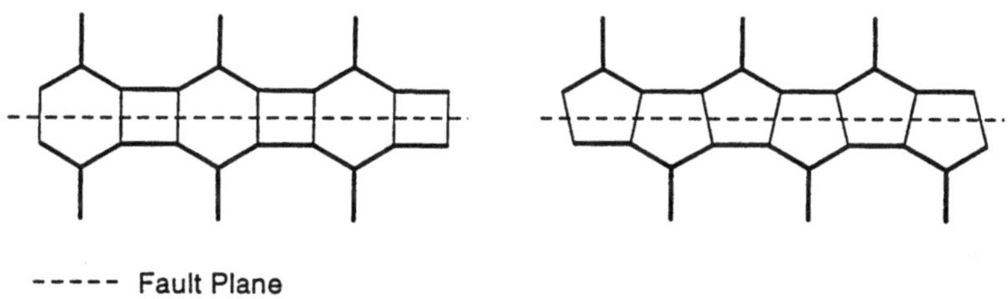

Fig. 6. Schematic diagram illustrating faulting

7 Loop Configurations and Coordination Sequences

The loop configuration is a simple 2-dimensional graph showing how many 3- or 4-membered rings a given T-atom is involved in. Figure 7 shows all observed loop configurations and their frequency of occurrence [2]. Sato [13] used the

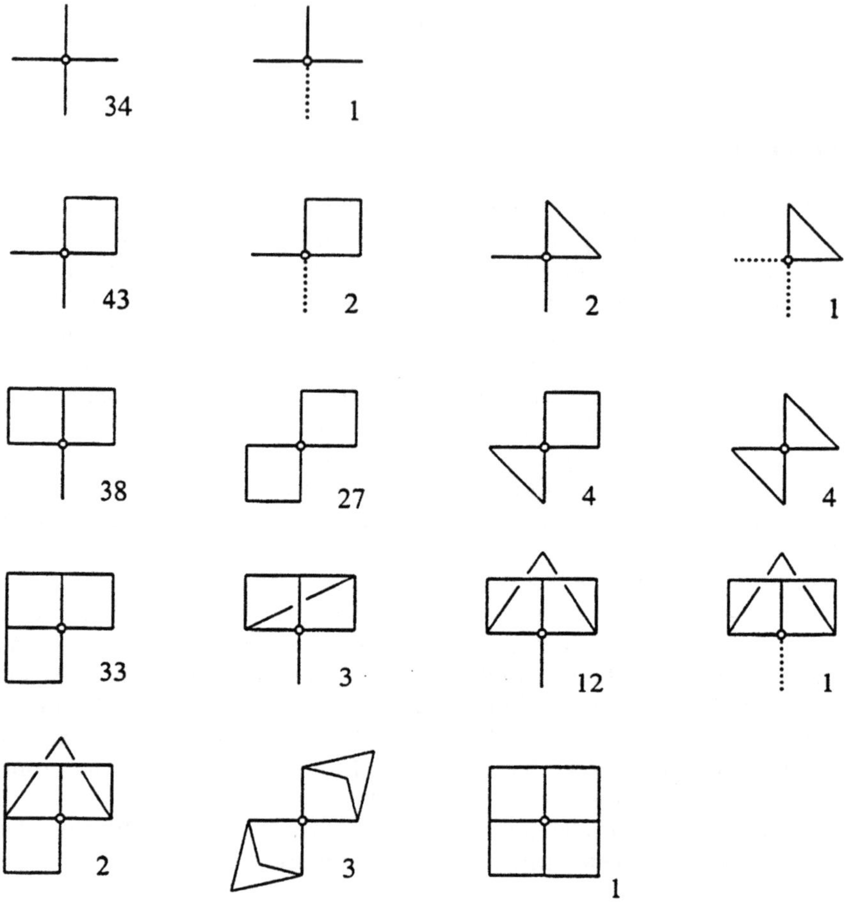

Fig. 7. Loop configurations and the frequency of their occurrence (adapted from the ATLAS)

term "second coordination networks". Loop configurations serve to characterize the immediate surrounding of the T-atoms in a zeolite framework and provide useful at-a-glance information particularly in the case of complex networks. They certainly supply information on the smallest rings present (dealt with in Sect. 4). Their use for classification purposes has been investigated, but no convincing scheme has yet been developed. More important are rules based on loop configurations which allow some prediction as to the likelihood of a hypothetical framework [14]. When considering Fig. 7 it is quite apparent that some loop configurations are very frequent while others are rare, and some (not shown but easily deduced) have never been observed. Loop configurations can to some extent also be correlated to the type of compositional system as shown in Table 3.

The concept of coordination sequences (CS) is, in a way, comparable to that of loop configurations, only the latter relates to local parts of a tetrahedral net, whereas the former comprises the full topology of the 4-connected net. In a zeo-

Table 3. Topological features of different zeolite-type frameworks

Framework composition	Prevailing loop configurations
High silica zeolites	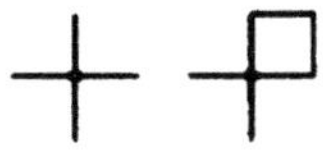
High alumina zeolitzes	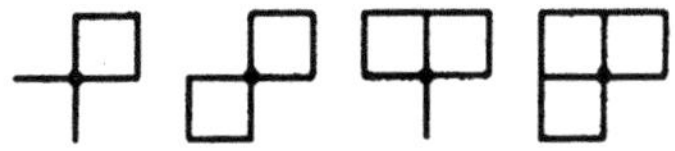
Aluminophosphates	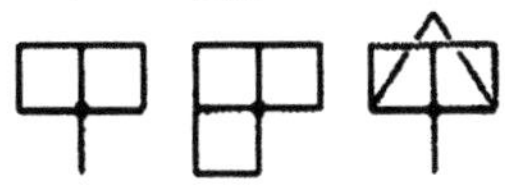

lite framework, each T-atom is connected to $N_1=4$ neighboring T-atoms through O-bridges. These neighboring T-atoms are then linked in the same manner to N_2 T-atoms in the next shell. The latter are connected with N_3 T-atoms etc., taking care to include T-atoms only once. In this way a CS can be determined for each T-atom of the 4-connected net of T-atoms. It follows that:

$$N_0=1 \quad N_1 \leq 4 \quad N_2 \leq 12 \quad N_3 \leq 36 \ldots \qquad N_k \leq 4 \cdot 3^{k-1}$$

Coordination sequences have been listed in the Atlas from N_1 up to N_{10} for each topologically distinct T-atom in the framework structure. A useful application of coordination sequences is the possibility of carrying out a computer search in order to find out whether or not a new topology has been observed before. In this way, isotypic frameworks can be easily recognized even when the crystallographic data differ considerably. Another interesting application of coordination sequences is the calculation of topological density (TD) values. The TD is in principle the total population of T-atoms up to a certain shell k. Unlike the FD the TD-values do not depend on unit cell volumes which can change on composition. As might be expected, the CS is a periodic function. This has been established for all observed framework topologies [15].

8
Framework Conformation

Non-rigid frameworks include those showing significant variations in FD. In many cases configurations also lend themselves to problems of conformation. Significant examples are apertures determining the size and shape of the free opening of a molecular sieve structure. Customary values are frequently given or assumed for the free diameters of channels with 8-, 10- and 12-ring apertures. It should be clear from the data in the Atlas that the crystallographic free diameter for a particular ring can vary greatly. The following example referring to 10-rings is representative

ZSM-5	5.3 × 5.6 Å
Heulandite	3.0 × 7.6 Å

The conformation of a ring is determined primarily by the connectivity and to a lesser extent by extra-framework constituents.

A number of zeolite frameworks are also known to undergo marked conformational changes. A good example is zeolite RHO [16, 17]. The framework structure of the desolvated NH_4-form exhibits considerable distortion as shown in Fig. 8, whereas the structures of the as-synthesized (Na,Cs)-form and the H-form are close to ideal. The changes in conformation of this RHO-framework are very pronounced though the desolvated NH_4-form is still cubic. This noteworthy example shows that cubic symmetry does not imply rigidity of the framework as is sometimes assumed.

Another enlightening example of framework conformation was provided by NaP1 which for quite some time was thought to be cubic. It was even assumed to have a separate ST in the earlier literature before the structure was solved [18]. The structure determination based on powder data showed it to be an isotype of GIS, the topology of which is clearly non-cubic. As shown in Fig. 9 natural gismondine (the type material) and NaP1 are isotypic but differ considerably in conformation. The difference is in fact so pronounced that the respective X-ray powder diffraction patterns (Fig. 10) would seem to indicate two different structure types. This is a good example showing that the identification of a ST of a zeolite-type material is not always a matter of routine.

On the other hand, it should be noted that conformational transformations do not necessarily manifest themselves directly by distinct shifts in the diffraction patterns. The shifts are normally no larger than those observed upon ion

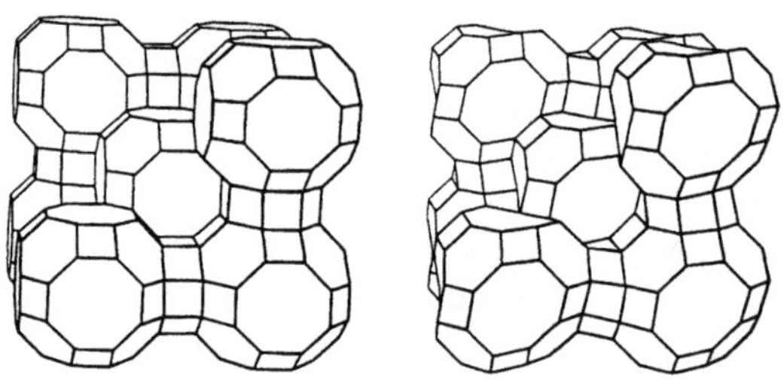

Fig. 8. Conformational changes in RHO. *Left*: H-form ($Im\bar{3}m$), *right*: desolvated NH_4-form ($I\bar{4}3m$)

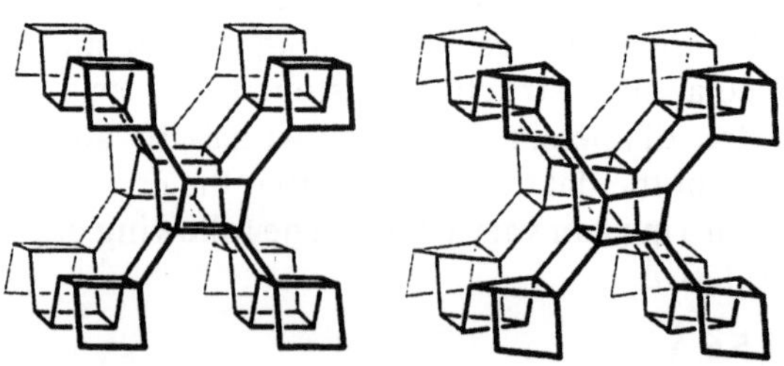

Fig. 9. Conformations of GIS-type materials. *Left*: gismondine ($P2_1/c$), *right*: NaP1 ($I\bar{4}$)

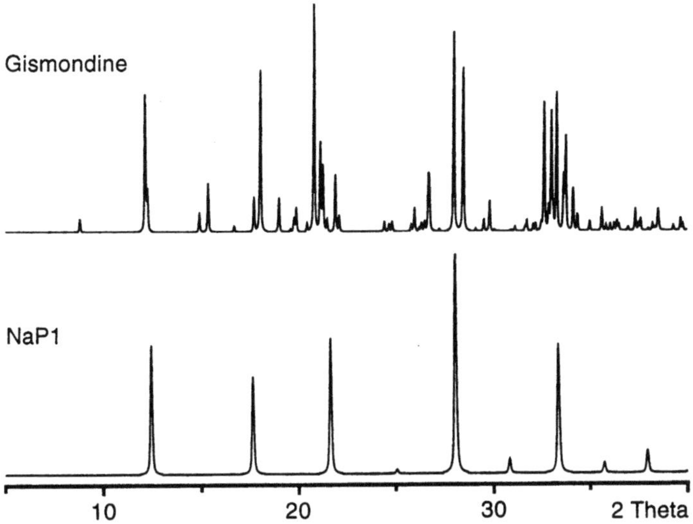

Fig. 10. XRD powder patterns of two GIS-type materials (*top*: gismondine, *bottom*: NaP1)

exchange or sorption. A shift in a diffraction pattern is thus not a clear indication of a conformational change. Therefore, changes in the free pore openings, due to changes in the framework conformation, cannot be detected directly by a diffraction experiment. A complete structure refinement must be performed to establish such changes.

Furthermore, it should be remembered that crystallographic free pore openings generally refer to the structure at room temperature. Possible changes in conformation have to be taken into account following chemical modification and/or heat treatment. Regrettably, only a limited amount of such structural information is available.

9 Are Zeolite Framework Structures Predictable?

Some 10% or more of the presently known framework structures were predicted well before they were actually observed. Hypothetical zeolite structures date back to at least the 1950s. Compilations of hypothetical frameworks have been set up by several investigators in the field ever since. By far the most extensive enumeration has been established by J.V. Smith (see for example [19]). Early examples of postulated zeolite-like framework structures include EMT (the hexagonal counterpart of FAU predicted by several investigators in the early 1960s but discovered only in 1990 [20]), DAC and RHO. In all these examples, the predicted structures were closely related to known framework structures (i.e. FAU, MOR and KFI, respectively).

The number of theoretical frameworks exceeds that of the observed ones by at least two orders of magnitude. The number is certainly very large but not infinite as has been stated repeatedly (cf. Sect. 5). FPs in known structures provide a simple basis for postulating hypothetical frameworks as already mentioned in Sect. 6. The result is typically a series of theoretical networks. Based on additional FPs found in some of these theoretical networks more of these nets can be generated in a systematic manner. This is only one of several procedures. Apart from a visual approach, computer methods have also been developed to generate large numbers of 4-connected nets making use of graph theory [21]. It should be noted, however, that a mere enumeration of theoretically possible connectivities includes a fair proportion of unlikely configurations. Theoretical frameworks should at least be tested using loop configurations and other means to assess their likelihood [14]. Further tests of postulated topologies should include distance least squares refinement (DLS) [22], a method specifically developed to optimize the geometry of zeolite structures. DLS was recognized quite early [1, 23] as a simple method to exclude structures containing unlikely bond distances and angles.

Since ab intio structure solution from powder data has made significant progress in recent years, the use of hypothetical frameworks for the solutions of new zeolite structures has lost some of its former importance. However, theoretical frameworks are still very attractive as a source of ideas for the scientist trying to synthesize new porous materials using various techniques and templating agents.

Acknowledgements. We thank Lynne B. McCusker for many helpful discussions and for critically reading the manuscript.

References

References to particular structures can be found in the Atlas [2, 3] and are not listed.

1. Meier WM, Villiger H (1969) Z. Kristallogr. 129: 411
2. Meier WM, Olson DH, Baerlocher Ch (1996) Atlas of zeolite structure types, 4th edn. Elsevier, New York. Published on behalf of the IZA Structure Commission
3. Baerlocher Ch, McCusker LB Atlas updates on http://www.iza-sc.ethz.ch/IZA-SC/
4. Bialek R, Meier WM, Davis M, Annen MJ (1991) Zeolites 11: 438
5. Strunz H, Tennyson Ch (1982) Mineralogische Tabellen, 8. Auflage. Akademische Verlagsgesellschaft Geet & Portig K.-G., Leipzig; Deer WA, Howie RA, Zussman J (1992) An introduction to the rock-forming minerals, 2nd edn. Longman, London
6. Brunner GO, Meier WM (1989) Nature 337: 146
7. Meier WM (1989) Zeolites and zeolite-like materials. In: Murakami Y, Iijima A, Ward JW (eds) Proc. 7th Int. Zeolite Conf., Tokyo. Kodansha, Tokyo. p 13
8. Meier WM, Olson DH (1992) Atlas of zeolite structure types, 3rd edn. Butterworth-Heinemann, London
9. Smith JV, Rinaldi F (1962) Miner Mag 33: 202
10. Breck DW (1974) Zeolite molecular sieves. Wiley, New York, pp 45–132
11. Gottardi G, Galli E (1985) Natural zeolites. Springer, Berlin, Heidelberg, New York, Tokyo, pp 4–33
12. Treacy MMJ, Vaughan DEW, Strohmaier KG, Newsam JM (1996) Proc R Soc London A 452: 813
13. Sato M (1984) Framework topology and systematic derivation of zeolite structures. In: Olson D, Bisio A (eds) Proc. 6th Int. Zeolite Conf., Reno. Butterworths, Guildford, p 851

14. Brunner GO (1990) Zeolites 10: 612
15. Grosse-Kunstleve RW, Brunner GO, Sloane NJA (1996) Acta Cryst A52: 879
16. McCusker LB (1984) Zeolites 4: 51
17. Baur WH, Fischer RX, Shannon RD (1988) Relations and correlations in zeolite rho and computer simulations of its crystal structure. In: Grobet PJ, Mortier WJ, Vansant EF, Schulz-Ekloff G (eds) Innovation in Zeolite Material Science, Proceedings of an International Symposium, Nieuwpoort (Belgium). Elsevier, Amsterdam. p 281
18. Baerlocher Ch, Meier WM (1972) Z Kristallogr 135: 339
19. Han S, Smith JV (1994) Acta Cryst A50: 302; Andries KJ, Smith JV (1994) Acta Cryst A50: 317
20. Delprato F, Delmotte L, Guth JL, Huve L (1990) Zeolites 10: 546
21. Sato M, Uehare H (1997) Stud Surf Sci Cat 105: 2299
22. Baerlocher Ch, Hepp A, Meier WM (1978) DLS-76, a program for the simulation of crystal structures by geometric refinement. (Inst. f. Kristallographie und Petrographie, ETH, Zürich)
23. Baur WH (1971) Nature 233: 135 and 447 (editorial)

Manufacture and Use of Zeolites for Adsorption Processes

A. Pfenninger

CU Chemie Uetikon AG, Zeochem Division, CH-8707 Uetikon, Switzerland;
e-mail: pfenninger.armin@zeochem.ch

1 **The Zeolite Market** 164

1.1 Major Zeolite Uses 164
1.2 The Role of Natural Zeolites 166

2 **Industrial Synthesis of Zeolites** 166

2.1 General Scheme of Zeolite Synthesis 167
2.2 Starting Materials 169
2.2.1 Silica Sources 169
2.2.2 Alumina Sources 170
2.2.3 Alkali Cation/Template 170
2.2.4 Hydroxide 170
2.3 Development of a Commercial Process 170
2.3.1 Reaction Parameters 172
2.3.2 Scale-Up Procedure 172
2.4 Important Synthetic Zeolites for Adsorption Processes 174
2.4.1 Zeolite A 174
2.4.2 Zeolite X 177
2.4.3 Zeolite ZSM-5 179

3 **Forming Processes** 180

3.1 The Ideal Binder System 181
3.2 Binderless Products 183

4 **Characterization of Zeolites for Adsorption Processes** 183

4.1 X-Ray Powder Diffraction 184
4.2 Elemental Analysis 185
4.3 Scanning Electron Microscopy 186
4.4 Bulk Density 188
4.5 Dust Formation 188
4.6 Attrition of Agglomerated Products 189
4.7 Bead Size Distribution 189
4.8 Adsorption Characteristics 190
4.9 Zeolite Stability 191

Molecular Sieves, Vol. 2

5 **Use of Zeolites in Miscellaneous Adsorption Processes** 193
5.1 Insulated Glass Industry 193
5.2 Drying and Purification of Air and Petrochemicals 195
5.3 Separation of Hydrocarbons 196
5.4 Separation of Gases 197

References 197

1 The Zeolite Market

The zeolite group of minerals was discovered in 1756 by the Swedish mineralogist Baron Cronstedt, who described several varieties occurring as well-formed crystals. He named them from the Greek "zeos" and "lithos", meaning "boiling stones", in allusion to their peculiar frothing characteristics when heated in an open test tube.

For the next 150 years not many investigations were done on zeolites. Beautiful specimens have been found over the years, but they were regarded as a curiosity of nature, and zeolites did not awake the interest of professionals. It was in the 1920s and 1930s when larger deposits of natural zeolites were found in the western part of the United States. The majority of the professional world remained unaware of these reports and occurrences. Only in the 1950s when these sediments were studied in more detail, especially by means of X-ray diffraction, was it discovered that these materials contained up to 90% of a single, well-defined zeolite mineral. Thick tuffs of clinoptilolite were found in Japan and large deposits of heulandite were detected in New Zealand. In the meantime some 40 different natural zeolite forms are known and well characterized. For a full historical perspective see [1].

In the middle of this century the first synthetic counterparts were prepared, and the industry showed an immediate interest in the zeolites A and X. From this starting position a huge market has developed over the last 50 years.

1.1 Major Zeolite Uses

A zeolite is a real stoichiometric inorganic polymer, each aluminum atom inducing one negative charge in the framework which requires a positive counterion. Zeolites became very attractive materials for many applications for the following reasons:

- zeolites are microporous materials with uniform pore dimensions;
- certain molecules may enter these pores, while others are rejected – this leads to selective adsorption processes;
- zeolites have very good ion exchange properties;
- zeolites may become strong Brønsted acids after ion exchange with an acid and, therefore, are good catalysts;
- zeolites are thermally very stable and can be used at high temperatures.

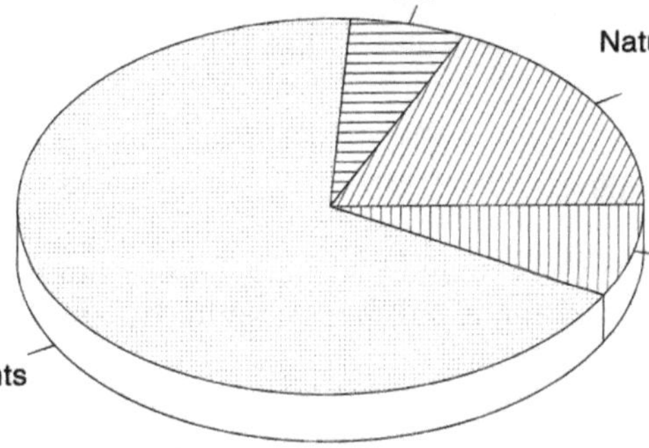

Fig. 1. Zeolite market volume

Total world-wide zeolite usage has become very important and is approaching 1.6 million tonnes per annum, zeolite A being the major part amounting to 1.1 million tonnes per year [2]. The segmentation of the market is shown in Fig. 1 [3–5].

The world demand for zeolite A is mainly by the detergent industry. In modern detergents, zeolite A is the major component, amounting to about 40% of the total weight. Each of Europe's twelve most important detergent zeolite producers has a capacity of 30,000 tonnes per annum or more [2].

In the field of adsorption and desiccation, it is again zeolite A which plays a major role. It is mainly used for removal of moisture and foreign substances from a gas or liquid mixture. One important application is as a desiccating agent in sealed insulation glass, which is composed of two or three panes of glass with a layer of dry air in between for heat insulation. In Europe and North America, use of sealed double or triple glazing has become common practice, and the demand for molecular sieves in that field is about 20,000 tonnes per annum. In Japan, use of sealed insulation glasses is limited to the northern districts such as Hokkaido and Tohoku, but their use is now increasing throughout the country [3]. Zeolites are also frequently used to dry various types of gases such as natural gas, ethylene, propylene, and air. After saturation with water vapor the zeolite is thermally regenerated. This type of process is known as thermal swing adsorption (TSA).

The separation of bulk chemicals has been carried out with zeolites for almost 40 years, the separation of *n*-paraffins and iso-paraffins being one of the first applications in this field [1]. The 1980s brought a huge number of pressure swing adsorption (PSA) processes to separate bulk gas mixtures, mainly air and reformer gases. This field of application is still a fast growing one.

In the catalytic application field, more than 90% of zeolites are used for fluid catalytic cracking (FCC) applications and in the hydrocracking market. The Y-type zeolites are typically used for these processes. FCC is an example where zeolites have had a large impact on both the crude oil economy and the cracker unit economy. Zeolite application made it possible to increase catalyst activity.

At a world-wide use of more than 8×10^6 tonnes of oil per day, a 1% gasoline yield increase represents an extremely high amount. Over the last two decades the yield improvements were well above 10% [5].

1.2
The Role of Natural Zeolites

The natural zeolite market is small in Western Europe but quite large in Japan. Natural zeolites are used for a number of quite different applications as is shown in the following paragraph.

Clinoptilolite may be used as a soft, high-brightness additive to paper. The material is first ground into fine particles and then cyclone-classified. The same material may also be used as a filler in the paper industry. It is again clinoptilolite which is used as a selective ion exchange agent, as it selectively removes cesium and strontium ions in the presence of high concentrations of competing ions. This process is especially useful for removing or concentrating and isolating radioactive species from waste waters generated by nuclear installations. In the early 1970s, natural, acid-washed mordenite had been used for the production of 90% pure oxygen. The oxygen was produced on-site for steel-making. Clinoptilolite was also studied for this process, although the nitrogen adsorption capacities of the samples tested are somewhat lower than those of mordenite. These zeolites have now been replaced by synthetic zeolites, which show a much higher performance than the natural zeolites.

Natural zeolites have also been in use as dietary supplements for swine and poultry. Up to 10 wt% of clinoptilolite and mordenite were added to the normal protein diet of pigs and chickens, with significant increases in the final weight of the animals and a concomitant reduction in the amount and cost of the feed. Zeolites apparently adsorb certain of the nutrients in the feed and retain them for longer periods in the animal digestive system, thus allowing a more efficient use of the feed and consequent decrease in the costs. Furthermore, the zeolite is able to bind the ammonia formed and thus helps to reduce odor emissions.

Another new application is found in municipal waste water treatment plants. Natural zeolites are able to bind significant amounts of ammonia, helping to remove nitrogen from these streams.

Finally, it should be mentioned that natural zeolites may be used as additives for fertilizers and for soil conditioning. It is believed that they may retain desirable cations in the soil for a longer period of time. Their high adsorption capacity may prove to be useful for carriers of pesticides and fungicides.

2
Industrial Synthesis of Zeolites

By definition, zeolites are crystalline, hydrated aluminosilicates of alkali and alkaline earth cations, having three-dimensional silicate structures. They are characterized by the ability to lose and gain water reversibly and to exchange constituent cations, both without major changes of structure. The framework consists of an arrangement of SiO_4 and AlO_4 tetrahedra in a manner which results in a remarkably open structure. The crystal structure of each member of the zeolite group is unique as shown by X-ray diffraction patterns. However, all of them are characterized by networks of channels or pores, leading to sizable central cavities. In the hydrated form the cavities are filled with alkali or alkaline earth cations which are surrounded by water molecules. The water may be removed by heating to about 350 °C, the cations then becoming coordinated with oxygen on the inner surface of the cavities. The channel dimensions of many zeolite species are large enough to permit molecules in the sub-nanometer range diameter to enter and become adsorbed in the vacant cavities.

In the 1950s a number of commercially significant zeolites, namely types A, X, and Y, were discovered [1, 6–8]. In 1954 Union Carbide Corporation commercialized synthetic zeolites as a new class of industrial materials for separation and purification purposes. The earliest application included the drying of refrigerants and natural gas. In 1959 the first process was introduced for normal- and iso-paraffin separation. At that time the term "molecular sieve" became quite popular. The 1980s saw major growth in the use of PSA for the production of oxygen, nitrogen, and hydrogen. Processes for the purification of gasoline oxygenate additives were also introduced at that time. All the early molecular sieves had a low SiO_2/Al_2O_3 ratio. Even today the major part of the molecular sieves used industrially for adsorption processes are still those with a low SiO_2/Al_2O_3 ratio. The low-silica zeolites are hydrophilic with a relatively weak acidity compared to the high-silica zeolites. The cation concentration in this group of zeolites is extremely high, and therefore they show a high ion exchange capacity. The hydrophilic low-silica zeolites are extremely well suited for removing water from organic compounds. In contrast, the hydrophobic high-silica zeolites and silica molecular sieves can remove and recover organic compounds from water streams and carry out separations in the presence of water.

2.1 General Scheme of Zeolite Synthesis

The earlier methods all involve hydrothermal crystallization of a reactive alkali metal aluminosilicate gel at low temperature and low pressure. The synthesis mechanism is generally described as involving solution-mediated crystallization of the homogeneous gel at a relatively high pH value. In the early alkali aluminosilicate synthesis of the low-silica zeolites, it was proposed that the hydrated alkali cation acts as a "template" – a forming agent – in the formation of zeolite structural sub-units. Water and hydroxide ions were the mineralizing agents. Crystallization of the zeolite was performed at a temperature close to 100 °C. The synthesis was varied in due course, raw materials changed, composition and reaction conditions were varied, but the principal lay-out of the reaction is still valid (cf. Fig. 2) [1].

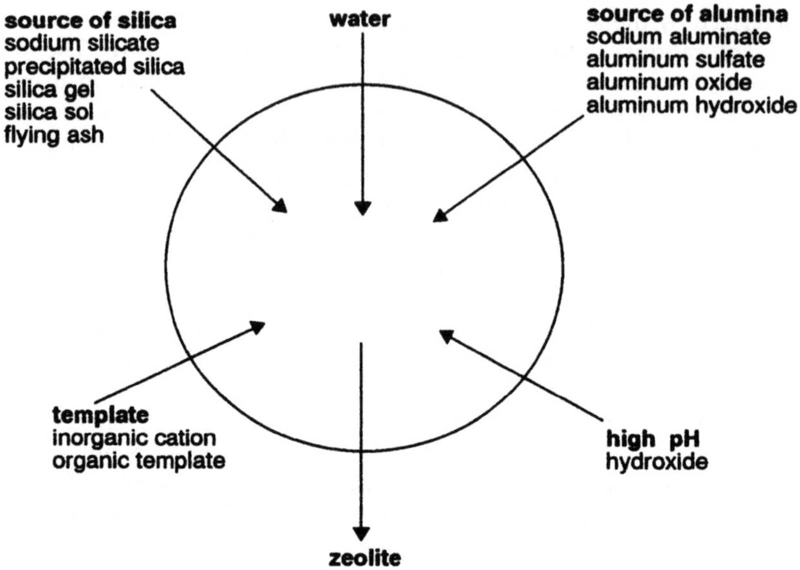

Fig. 2. Principle of zeolite synthesis

Each of the reagents used in the zeolite synthesis has its particular role. The synthesis is normally performed in water as a solvent. This is identical with the geological formation of natural zeolites. Water is not only a solvent to dissolve the starting materials but also a guest molecule as it hydrates the cations and thus leads to the hydrated template. The silica source acts as a primary building unit of the whole zeolite framework, each silica being connected with four oxygen atoms to form a tetrahedral network. Alumina acts as the second primary building unit and is connected to four oxygen atoms like silica, again contributing to the tetrahedral network. As aluminum is trivalent, it becomes the origin of the negative charge of the zeolite framework. According to Löwenstein's rule, it is not possible to get two alumina centers as neighbors in the zeolite framework, as the two negative charges would destabilize the whole system [9]. Due to this fact, the SiO_2/Al_2O_3 ratio always is ≥ 2. Hydroxide acts as a mineralizing agent and has a big influence on the result of the zeolite synthesis. The kinetics of zeolite formation is controlled partly by the hydroxide, but the size of the zeolite crystals is also influenced by the basic compound. The alkaline or the alkaline earth cation will counterbalance the negative charge of the alumina built into the framework and acts as the template for the structure-forming process. The main functions of the reagents are summarized in Fig. 3.

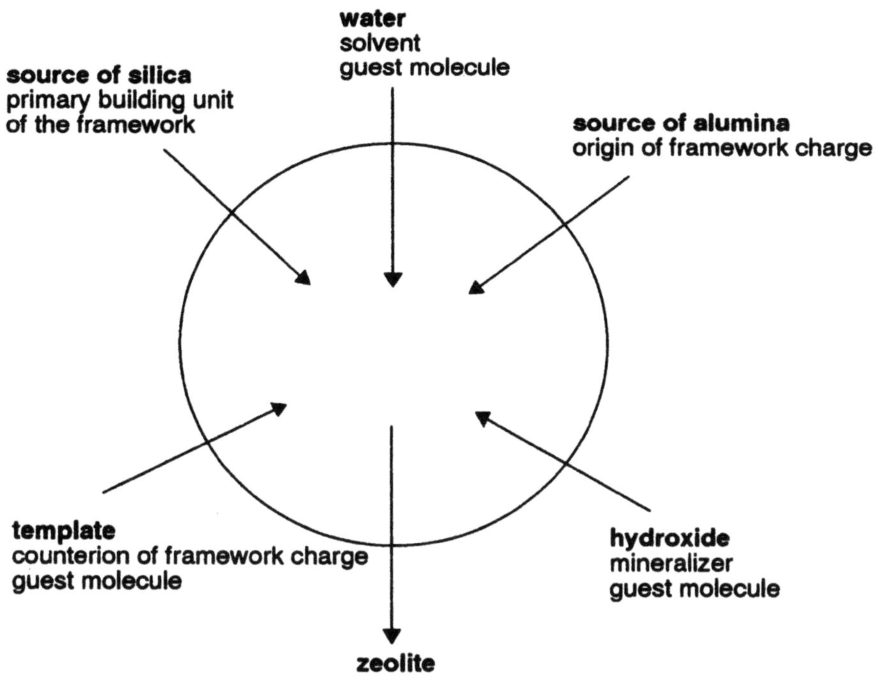

Fig. 3. Functions of the reagents

2.2 Starting Materials

2.2.1 *Silica Sources*

Depending on a particular synthesis, one silica source might be more favorable than another [10]. Quite a number of factors will influence the choice of a silica type. Pyrogenic silica is readily dissolved in the reaction mixture due to the small particle size. The rate of dissolution is an important parameter with a major influence on the rate of nucleation and crystallization meaning that another silica source that dissolves much more slowly may lead to a lower rate of zeolite formation or eventually to a different type of zeolite. Also the degree and the type of impurities have to be considered carefully. Cations in particular may act as templates and thus lead to another zeolite structure. The tetraalkoxysilanes are available in very good quality and may be further purified through distillation, but they are expensive and produce 4 moles of an alcohol per mole of silicon. For industrial applications this is a high barrier to overcome, as the equipment is normally not set-up in an explosion-proof area.

The cost of raw material is an important consideration apart from the fundamental suitability of the materials. The water glass price is typically less than 0.5 US$ per kilogram, whereas pyrogenic silica grades may be ten times more expensive. For the industrially available molecular sieves, sodium water glass, sodium disilicate, and precipitated amorphous silicas are the preferred starting materials.

2.2.2
Alumina Sources

The same arguments hold as for silica sources. Quality, price, availability, safety in handling, and solubility have to be evaluated. For the cheaper molecular sieves, sodium aluminate or aluminum hydrate are the starting materials of choice. Sodium aluminate has to be stabilized with an excess of hydroxide ions, otherwise it will lead to a precipitation of aluminum hydroxide. This is a disadvantage as the mother liquor will contain an excess of hydroxide ions which have to be neutralized or recycled in the process.

2.2.3
Alkali Cation/Template

If possible an inorganic cation, mainly an alkaline cation, is used for the synthesis. Sodium hydroxide is used except for those cases in which the presence of sodium ions prevents the formation of the desired zeolite type. For the production of the high-silica zeolites, an organic template is frequently used. In the open accessible literature a huge variety of different organic molecules are described, charged and neutral molecules, quite often ammonium cations or amine compounds [10]. For both practical and cost reasons, in industry the quaternary ammonium cations are mainly used as templates. They are not volatile and therefore easy to handle, they do not smell, and may be recycled to some extent.

2.2.4
Hydroxide

Most zeolite syntheses are performed under basic reaction conditions, using hydroxide as a mineralizing agent. Fluoride would be an alternative but is not used on a commercial scale as it is very corrosive to the equipment used.

2.3
Development of a Commercial Process

To optimize a zeolite synthesis on a commercial scale, it is necessary to investigate each unit operation very carefully. Quite commonly the zeolite synthesis is divided into four steps (Table 1) [10]. At a relatively low temperature all the ingredients are weighed or metered and mixed in the reactor. The easiest way is to add them as solutions. The order of addition and the time required for the addition is very important and has to be optimized for each individual application, including the seeds if any are added. The starting mixture is stirred to get

Table 1. Sequence of events in zeolite synthesis

Temperature	Event
< 60 °C	Mixing of starting solutions Forming reaction mixture Gel formation
Heating	Gel rearrangement Dissociation of silicate Dissolution of gel
< 200 °C	Nucleation Crystallization
< 100 °C	Isolation of zeolite

a homogeneous gel. The pH value and the concentration used will influence strongly the composition of the gel formed, mainly the degree of polymerization which has an influence on the viscosity. The stirring time of the gel prior to heating leads to some rearrangement and aging. Without an aging step it is not easy to produce a proper X zeolite that is not contaminated with other zeolitic phases.

In the next step, the gel is heated to the desired temperature. At this stage of the process the gel will rearrange, dissolve, and eventually nucleation of the zeolite will start. A secondary effect may occur. The quaternary ammonium ions may start to degrade and decompose to a tertiary amine and to a volatile olefin. This may lead to an unwanted increase in pressure and to a decrease in pH, as the hydroxide ions are consumed in such a Hoffmann degradation.

At the reaction temperature crystallization will start. Species in solution rearrange from a continuous changing phase of monomers and clusters, e.g., polysilicates and aluminosilicates. These clusters form and disappear via condensation and hydrolysis processes. The clusters and cations will start to associate, and the process particles reach the critical minimum size and become stable. For zeolite A, this minimum dimension is about 1.0 nm. At this stage crystallization starts. The liquid phase is continuously present between the dissolving dense gel phase and the growing crystals, and the crystallization is solvent-mediated. The size and homogeneity of the formed crystals will depend very much on the set of parameters used [11].

Mostly, the product is isolated by centrifugation or filtration. This process is followed by washing with water to remove the mother liquor and especially the excess of hydroxide ions. This unit operation, usually done at slightly elevated temperatures, is crucial. The valuable zeolites are metastable products and may rearrange to more stable phases if not filtered and washed properly. The final process in the zeolite preparation is the drying or calcination procedure. This again is a delicate step as the metastable products may undergo hydrothermal changes if not properly washed and dried in thick layers where a high partial pressure of water vapor may build up. After the drying procedure, the void volume is free for different modifications and/or applications.

2.3.1
Reaction Parameters

A good industrial process should not only use cheap and readily available starting materials but should also be easy to run on a large scale. Heating and cooling a reaction mixture is a time-consuming and energy-intensive unit operation. In an ideal case, all of the unit operations mentioned above run at the same temperature. Furthermore, the reaction time should be as short as possible. In a well optimized system, the synthesis time for zeolite A or zeolite X does not exceed 3 h.

Zeolitic systems to be used for adsorption processes should be built up of small crystals (1–4 μm) to get reasonably fast kinetics. The crystal size is influenced through a number of parameters as shown in Table 2.

For zeolite A and zeolite ZSM-5, increasing the pH brings about an increase in the crystallization rate. The hydroxide ion is a strong mineralizer, bringing the reactants into solution. The higher the pH and thus the concentrations of the dissolved reactants, the more the rate of crystal growth of the zeolite is enhanced, the more nucleation centers are formed, and the smaller is the crystal size. It has been shown for many zeolites that raising synthesis temperatures within the zeolite synthesis field increases the crystal growth rate. In the case of zeolite A the crystal size does not change significantly, whereas in the case of ZSM-5 substantial variations in the crystal size are produced – the higher the temperature, the larger the crystals.

2.3.2
Scale-Up Procedure

In order to develop optimum production processes for existing and new products, development is normally done in several stages. The first step consists of

Table 2. Parameters which influence the crystal size

Parameter	Effects
pH variation	High pH (high concentration of hydroxyl ions) will produce small crystals Low pH (low concentration of hydroxyl ions) will produce large crystals
Na_2O/SiO_2 ratio	A low Na_2O/SiO_2 ratio leads to large crystals (nucleation is very slow) A high Na_2O/SiO_2 ratio leads to small crystals (nucleation is very fast)
Influence of seeds	Large amounts of seeds quite often lead to smaller crystals (especially for zeolite Y) Addition of seeds normally leads to shorter synthesis times
Aging of the gel	Aging of the gel favors nucleation in a short time and leads to small crystals
Stirring speed	Higher stirrer speed leads to smaller crystals

optimizing the synthesis on a laboratory scale. Preliminary experiments are done on a very small scale, usually in small reactors of 100 ml to 1 l. The reactor type is dependent on the reaction to be performed. A glass reactor is used if the reaction temperature is not going to exceed 100 °C. These reactors have the big advantage of being transparent so the experiment can easily be followed by visual inspection. The further advantage is that these reactors may be cleaned quite easily. At this stage of development, it is absolutely necessary to work with perfectly clean equipment in order to exclude cross contamination or unwanted seed effects. Unfortunately, the glass reactors are not resistant to the reaction conditions over a longer period.

If the reaction temperature is going to exceed 100 °C, a pressurized steel reactor is needed. Advantageously, the reactor is teflon-lined to exclude metallic effects from the autoclave's wall. The teflon liner is either new or is cleaned by boiling with 30% sodium hydroxide for a couple of hours to destroy all the crystalline contaminants.

Once the reaction works satisfactorily on a 100 ml scale, it will be performed on a scale of 5–10 l. The same precautions will be taken as for the initial step. A few new items are considered at this stage of development. The process has to be made simpler and sturdier. The influence of some impurities are studied at this stage. If the reactor for commercial scale production is not cleaned after each operation, some residues may build up at the reactor wall after some time. These residues in turn may act as seeds or, in a worst case, may help to form an unwanted zeolitic phase. The quality of the raw material may change over time. It is well known that the composition of water glass is very much dependent on the dissolution process and does not remain constant over a lengthy time period.

At this stage, the temperature limits of the reaction are studied. For instance, an organic template, a quaternary ammonium salt, may undergo degradation and build up a pressure that may become dangerous on a large scale.

Seeding techniques are studied to gain insight into the kinetics of the process and to optimize the crystal size. Too large crystals may show poor kinetics in the adsorption process or the catalytic reaction they are used for, while too small crystals are not easy to filter and may not be stable enough for the envisaged process.

Finally, ecological and economical aspects are studied. The composition of the mother liquor is carefully analyzed to discover whether the reaction mixture has to be adapted or whether some of the mother liquor may be recycled (Table 3).

Table 3. Parameters which lead to an optimized zeolite synthesis

Make the process simpler and sturdier
Study the influence of impurities
Evaluate temperature limits of the reaction
Check different seeding techniques
Optimize the crystal size
Study ecological and economical aspects

Fig. 4. Photograph of a semi-commercial plant for zeolite synthesis. *Left* is the stirred, pressurized reactor, *middle* is the pressurized reactor with heating coils, and *right* is the open reactor with heating coils and stirrer

The final optimization stage is carried out in a semi-commercial plant with reactors of a size of up to 1500 l. Open steel reactors with stirrers and heating devices (double jacket or heating coils) are adequate to run syntheses of zeolites such as the A type or the X type. A stainless steel reactor with a stirrer and a working pressure of up to 12 bar is needed to develop the high-silica zeolites that are synthesized with an organic template at elevated temperatures up to 160 °C. A closed steel reactor with a dense array of heating coils, which ensure a very low temperature differential within the synthesis batch, may be needed for the synthesis of zeolites that are not easily prepared under stirred conditions, for example zeolite L (Fig. 4).

The finished zeolite is separated from its mother liquor and thoroughly washed on a filter and finally dried in a muffle oven as shown in Fig. 5.

2.4
Important Synthetic Zeolites for Adsorption Processes

2.4.1
Zeolite A

Zeolite A was the first zeolite to be synthesized on a large scale, and it is still the working horse for a wide range of present adsorption applications. It is

Fig. 5. Muffle oven of the semi-commercial plant to dry and calcine the zeolites after synthesis and filtration

estimated that the world-wide consumption of zeolite A for adsorption and desiccant application is about 95,000 tonnes per annum [3].

Zeolite A is synthesized in the sodium form to give a pore opening of 0.38 nm. A simple ion exchange with potassium salt after the filtration and washing steps reduces the pore size to 0.30 nm. Similarly, ion exchange with calcium salt can increase the pore size to 0.43 nm. These seemingly small differences in pore opening have a crucial effect for many applications. The potassium, sodium, and calcium forms of zeolite A are usually referred to as zeolites 3 A, 4 A, and 5 A, respectively.

The process flow sheet (Fig. 6) illustrates the type of equipment and the raw materials used. Sodium water glass and sodium aluminate solutions are prepared in premix tanks and then metered into the reactor in the correct ratio at 50–70 °C. The gel is formed and aged at this temperature. To this mixture seed crystals may be added, and then the gel is heated to near the boiling point, crystallization gradually taking place at this temperature. The resulting slurry is filtered, washed, and then dried. For further processing the filter cake is ground to a fine powder. Part of this powder is calcined directly to yield a product which is used as a desiccant in the glazing and polymer industries.

The larger part is mixed with a natural clay binder and then shaped in large granulator wheels. Sieving and calcination at above 600 °C are the final steps before the beads are packed in airtight polyethylene bags inside either cardboard boxes, steel drums, or large bags.

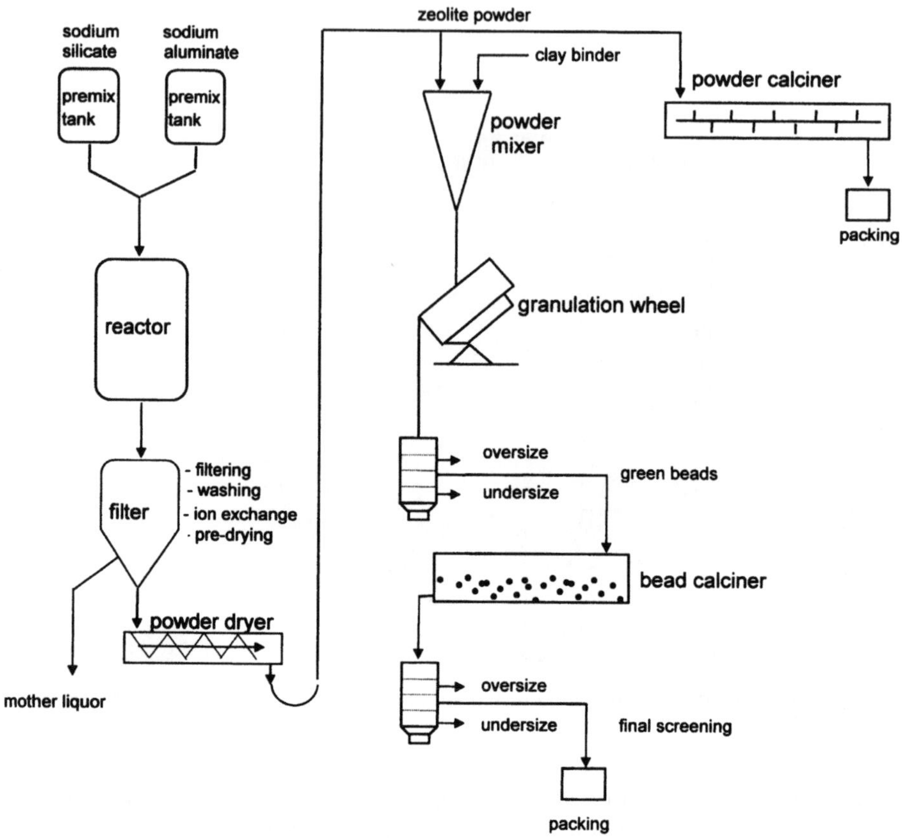

Fig. 6. Synthesis of zeolite A: process flow sheet

Theoretically, zeolite A has a SiO_2/Al_2O_3 ratio of exactly 2.0, the lowest figure possible according to Löwenstein's rule [9]. In reality, however, zeolite A has a SiO_2/Al_2O_3 ratio that fluctuates between 1.92 and 2.08, and it has been shown that the products with higher SiO_2/Al_2O_3 ratio tend to be more stable than those with a lower SiO_2/Al_2O_3 ratio.

In the production of zeolite A, the stoichiometry of the reaction mixture can be varied over a relatively wide range. However, efficient production requires optimization of the composition from the point of view of product quality, yield, and reaction time. One possible composition [12] is

$$2.04\ Na_2O{:}1.00\ Al_2O_3{:}1.75\ SiO_2{:}70\ H_2O$$

and may be composed of 1350 parts of sodium silicate (9.4% Na_2O; 28.4% SiO_2), 575 parts of aluminum trihydrate (65% Al_2O_3; 35% H_2O), 870 parts of sodium hydroxide (50% NaOH in water) and of 3175 parts of water. This relatively concentrated reaction mixture yields about 150 kg of zeolite per 1000 l of reactor volume.

The ion exchange of the filtered and washed zeolite crystals may be done at various stages. One possibility is to do the ion exchange directly on the belt filter where the zeolite was filtered and washed. In a subsequent step, potassium or calcium salt solutions are added in the required amounts to achieve the ion exchange. The salt solution is drained, and the filter cake is washed again with water. Such systems are in use, but they do not lead to a homogeneous ion exchange, and there is an ion gradient across the filter cake.

To achieve a homogeneous ion exchange, the filter cake is slurried in a tank after washing and the potassium or calcium salt is added. After exchange, equilibrium is achieved and the slurry is filtered and washed in the usual way. This process leads to a uniform ion exchange level.

Still another approach starts from the formed zeolite, which is put into an exchange column. The column is percolated with the salt solution to achieve the ion exchange. After the ion exchange is completed the column is washed with water and the zeolite activated in the usual way. Instead of a salt solution a hydroxide solution may be used to convert the kaolin binder into zeolite (see Sect. 3.2). This conversion takes place at elevated temperatures of 80–90 °C.

2.4.2
Zeolite X

The synthesis of zeolite X is similar to the one of zeolite A, the pore size being 0.74 nm instead of 0.38 nm. Exchange of the sodium with other cations is also possible, but the changes in the pore size have not gained the same importance in practical applications as for zeolite 4 A. However, the exchange of the sodium cation with calcium or lithium changes the electrostatic charge distribution and leads to different adsorption properties of the zeolite. The calcium exchanged zeolite A, and especially the lithium exchanged zeolite X, are used for the PSA separation of air [13, 14].

There are some significant differences compared with the zeolite A synthesis. Zeolite X is less stable than zeolite A and may easily be converted to zeolite P by prolonged synthesis time or too high synthesis temperature. With variation of the stirrer speed during synthesis from 150 to 350 rpm, the yield of zeolite X passes through a maximum followed by a slow decrease in which zeolite X is progressively converted to zeolite P [11]. Heating the reaction system to higher temperatures than usual may produce zeolite A instead of zeolite X. Upon heating the finished zeolite X at temperatures of between 100 and 600 °C in the presence of water vapor, the relative intensities of the lines in the X-ray pattern may be appreciably changed from those existing in the unactivated zeolite X patterns [6]. Zeolite X needs more hydroxide excess for crystallization as is seen from the crystallization fields in Figs. 7 and 8 [11]. This causes extra problems in achieving an economically and ecologically suitable process.

Another interesting point is that zeolite X is available with different SiO_2/Al_2O_3 ratios between 2.0 and 3.0. Within a system containing only sodium as cations, it is not easy to achieve the very low SiO_2/Al_2O_3 ratio of 2.0 [16]. However, adding potassium to the reaction system easily allows the synthesis of the

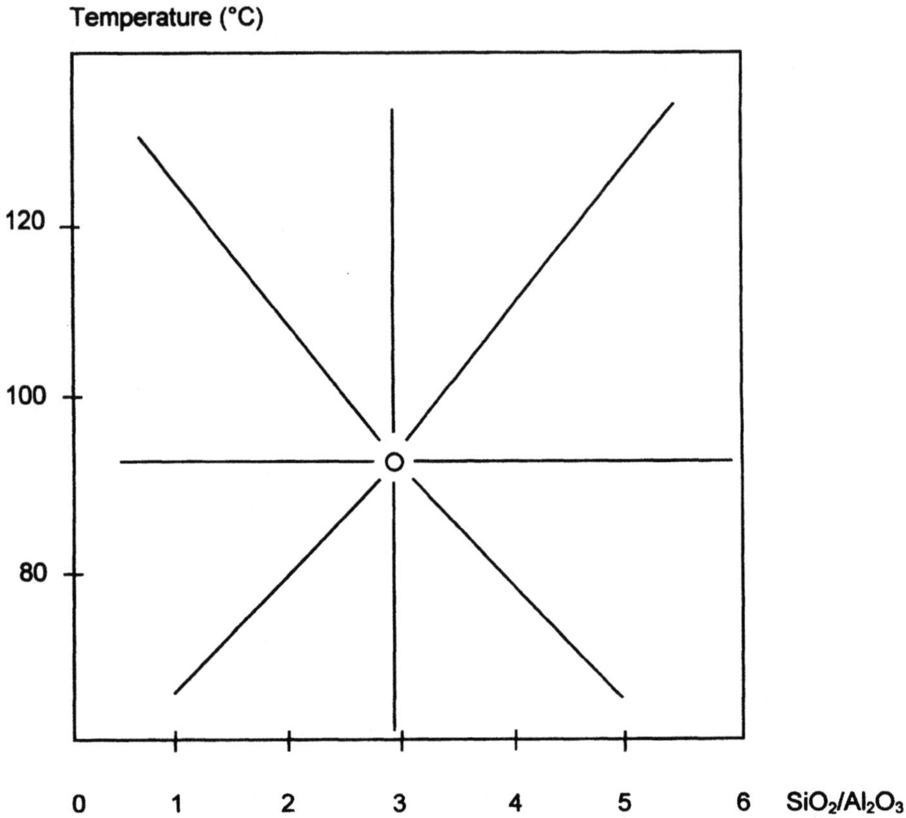

Fig. 7. Crystallization field for zeolite A (the excess amount of NaOH per Al_2O_3 is 100–300%)

low SiO_2/Al_2O_3 ratio zeolite X (LSX) [17]. The LSX is an interesting product because of higher density of cations which eventually leads to higher adsorption capacities [15, 18]. A further important characteristic of zeolite X is its ability to adsorb gases at low partial pressures. This property makes zeolite X useful in the removal of impurities from gas and liquid mixtures [6]. As an example, zeolite X may be used to remove carbon dioxide from a predried air stream.

A typical gel composition that may be used to synthesize zeolite X is [6]:

$$3.50\ Na_2O{:}1.00\ Al_2O_3{:}3.00\ SiO_2{:}144\ H_2O$$

Thus, mixing 2000 parts of sodium silicate (9.4% Na_2O; 28.4% SiO_2) and 500 parts of aluminum trihydrate (65% Al_2O_3; 35% H_2O) dissolved in 1600 parts of sodium hydroxide solution (50% in water) and diluted with 7687 parts of water will yield the above-mentioned gel formula and, after crystallization at about 80 °C, to zeolite X. A 1000 l reactor volume may yield about 90 kg of zeolite X.

In making the sodium form of zeolite X, representative reactants are silica gel, silicic acid, colloidal silica, or sodium silicate as silica sources. Typical alumina

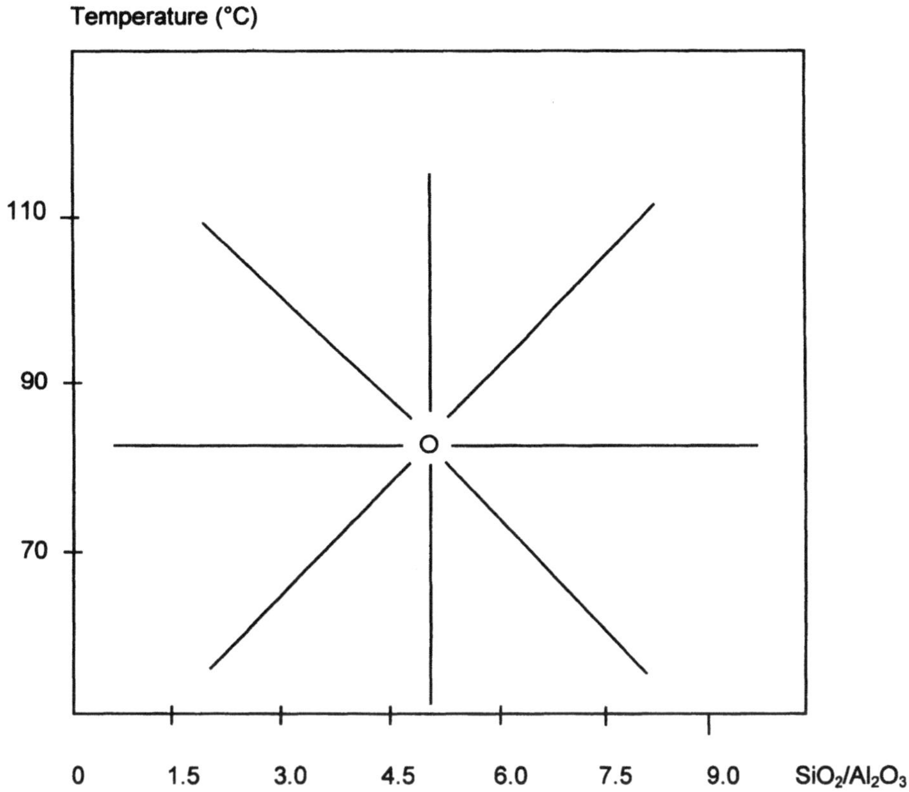

Fig. 8. Crystallization field for zeolite X (the excess amount of NaOH per Al_2O_3 is ≥300%)

sources are activated alumina, γ-alumina, alumina trihydrate, or sodium aluminate. Sodium hydroxide may supply the sodium ion and in addition assist in controlling the pH.

As mentioned above, sodium as a cation may be replaced by other cations through ion exchange. A preferred continuous method is to pack zeolite X into a series of vertical columns, each with suitable supports at the bottom. Subsequently, a water solution of a soluble salt of the cation to be introduced into the zeolite is passed through the bed. The flow is switched from the first bed to the second bed, as the zeolite in the first bed becomes ion-exchanged to the desired extent [6].

2.4.3
Zeolite ZSM-5

Zeolite ZSM-5, with a pore size of 0.54–0.56 nm, may be synthesized from different silica sources using synthesis procedures with and without organic templates. In the syntheses with templates, the organic molecules serve as building blocks around which the specific structures of the zeolites can be synthesized. The first criterion is the geometric size and shape of the molecule which has to match the crystal cavities to be formed. Further considerations include the stability against hydrolysis under synthesis conditions, the toxicity, and – last but not least – the cost of the template. In most cases the template used is a partially soluble amine or a tetrapropyl ammonium salt. After completion of the synthesis, which is normally performed at 130–160 °C and under autogenous pressure, the excess template has to be recirculated or recovered from the mother liquor and wash water. This proves rather difficult in most cases, resulting in either high cost for the recirculation operations or in high waste treatment costs plus the cost of new template for the next batch. As the SiO_2/Al_2O_3 ratio of ZSM-5 can vary over a wide range, ZSM-5 is not a single material but a family of zeolites. It was first synthesized by Mobil researchers [19]. For SiO_2/Al_2O_3 ratios up to about 50 the non-templated synthesis is preferred. The SiO_2/Al_2O_3 ratio can be adjusted over a very wide range by altering the composition of the synthesis gel; if templates are used, SiO_2/Al_2O_3 ratios from 30 to infinity (aluminum-free form of ZSM-5, silicalite-1) can be achieved [20]. The crystal suspension obtained after synthesis is worked up as described above for zeolite A. If an organic template was used, it must be burned off from the ZSM-5 pores by heating to above 500 °C in air.

ZSM-5 zeolite is a silica-rich zeolite and therefore highly hydrophobic. In the field of adsorption it is mainly used to pick up volatile organic compounds from a mixed gaseous or liquid stream.

3 Forming Processes

All the zeolites described in the previous sections are formed as fine powders, the crystal size being in the range 1–5 µm. These powders are used directly for special purposes only. One of these applications is found in the polyurethane industry. Zeolite A is added to the polyol component to adsorb the residual water. Water would react with the isocyanate part of the polymerization system and eventually form carbon dioxide which would lead to bubbles in the system. Another important application is related to the insulating glass industry. The aluminum profile filled with molecular sieve beads may be replaced by a porous silicone foam system containing zeolite A powder as adsorbent. With these polymer systems even better insulation values for the window panes are achieved than with the existing solutions.

For most of the other applications the zeolites are used in agglomerated form, either as cylindrical pellets (extrudates) or as spherical beads. These agglomerates must have high physical strength and attrition resistance and should not form any dust upon handling. Methods for forming the agglomerates include the addition of an inorganic powder, generally a clay. The mixture is thoroughly mixed and wetted with water and then extruded [21] or formed into beads [22] in a granulation wheel. The shaped material is then dried at 100 °C and subsequently calcined above 600 °C to convert the clay to the fixed binder. A dry purge gas should be passed through the furnace during the firing operation to minimize the loss of adsorptive capacity of the zeolite. The maximum temperature for the firing process is the highest temperature at which the molecular sieve is structurally stable. For zeolite X and zeolite A this temperature is typically 700 °C.

The kaolin-type clays pass through an irreversible phase transition at 600 °C, providing a product having maximum crush strength and attrition resistance. Attapulgite-type clays undergo an irreversible phase transition between about 425 °C and 525 °C [22]. The most important clays used for binding zeolites without substantially altering the adsorptive properties of the molecular sieves are kaolin, attapulgite, sepiolite, and bentonite.

3.1
The Ideal Binder System

Different clay types are in use for the forming process, depending on the application field. The binder has to fulfill several requirements. As mentioned above, the agglomerates must have a high physical strength. This may be achieved by using a large amount of binder. The amount of binder used is a trade-off between crush strength and adsorption capacity of the final system, as the binder is an inert material and does not contribute to the adsorption (Fig. 9). Normal-

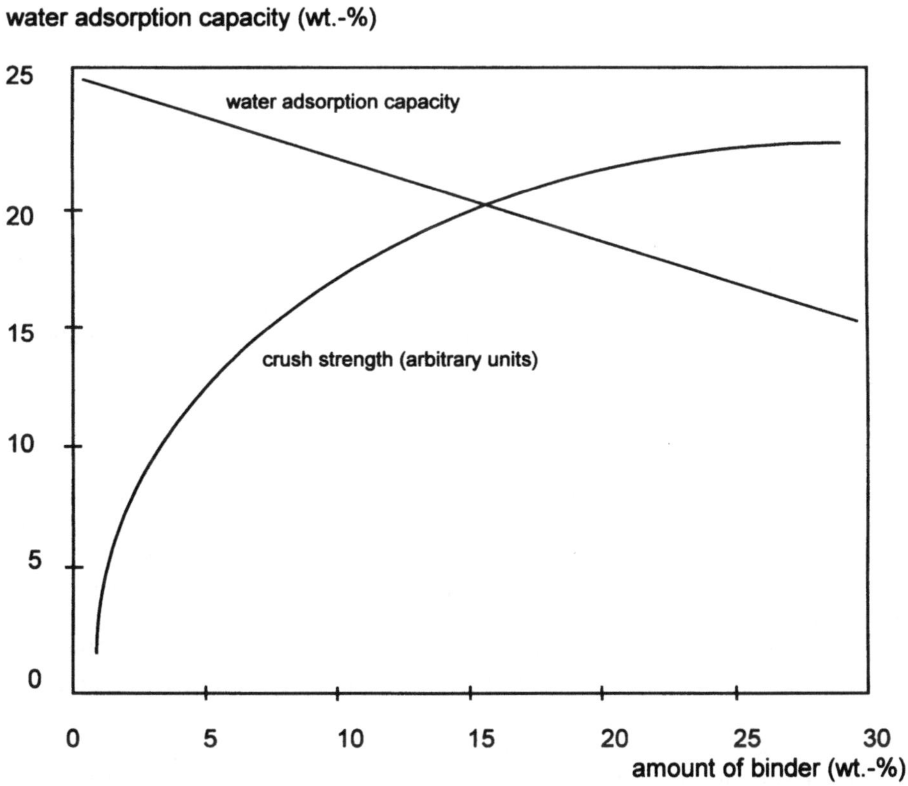

Fig. 9. Influence of the amount of binder on the crush strength and the water adsorption capacity [23]

ly, the systems contain 15–25 wt% of binder. A second important point is again related with the adsorption process. The diffusion characteristics must not interfere with the binder. The binder component must maintain a high macroporosity which does not increase the diffusion resistance. A very dense extrudate, produced under strong shearing forces, may have a high density and a high crush strength but low macroporosity. Therefore, the system has to be optimized to achieve an extrudate of high density, with maximum crush strength, but minimum restriction of diffusion.

Some applications are based solely on the equilibrium adsorption capacity of the system with essentially infinite time being allowed for adsorption. Drying of refrigerants in a cooling system or drying the air in an insulating double pane glass system are examples of such processes. For these systems the kinetics of diffusion are not as important as a high adsorption capacity for the agglomerated system.

Kaolin and attapulgite clays are used on a frequent basis but others such as sepiolite and bentonite clays may also be used for special applications. To choose the best binder one needs to know their characteristics fairly precisely.

The medium particle size and the particle size distribution of a binder should be similar to the zeolitic powder. Only in these cases may high binding forces develop during forming and calcining of the agglomerates. The bulk densities of the various clays vary remarkably, the lower end being at 150 g/l, the upper end at 1100 g/l. Choosing an appropriate binder may help to adjust the bulk density of the finished product to the required values. Kaolin clays are generally preferred as the binding material for extruded pellets. However, while spheres may be prepared, the fired spheres have limited strength. On the other hand, attapulgite-bonded spheres have superior fired strength to the attapulgite-bonded extrudates. It is also to be noted that the attapulgite spheres are generally much more attrition-resistant than the kaolin-bonded extrudates [22].

With some clay binders such as montmorillonite and bentonite the crush strength of the non-calcined and the fired extrudates is good when the material is dry but is seriously impaired by water.

Under optimized conditions, for each binder system extrudates were produced from 20% binder and 80% zeolite A powder. After calcination at 650 °C the apparent density and the crush strength of the extrudates were determined. The density ranged from 820 to 1230 g/l and the crush strength from 42 N to 125 N for 1/8" extrudates. Interestingly, the BET surface areas vary a lot from one binder type to another with values between 10 m^2/g and 150 m^2/g. Unfortunately these figures are not easily correlated with the macroporosity and the performance of the formed system.

Last but not least, the price of the binder is also an important factor. Obviously it depends on the purity of the binder, the availability, and the type. Some of the main data are summarized in Table 4.

In the patent literature it is also claimed that the addition of organic additives to the mixture of zeolite and binder will lead to improved crush strength of the extrudates. Addition of 2 parts of lignosulfonate to a mixture of 80 parts of zeolite X and 20 parts of a kaolin clay led to an improvement of the average crush

Table 4. Characteristic data of binders suitable for zeolites

Binder type	Medium particle size (μm)	BET surface of binder (m^2/g)	Bulk density of binder (g/l)	Bulk density of extrudates (g/l)	Crush strength of extrudates (N)
Kaolin	4–22	25–43	730–1150	990–1250	49–126
Attapulgite	7–12	82–112	350–770	850–930	42–81
Sepiolite	9–14	113–148	150–550	830–910	50–82
Bentonite	10	12	870	1050	95

strength by a factor of 3.8 [21]. This effect was not found if kaolin was replaced by attapulgite or bentonite.

3.2 Binderless Products

Several methods deal with the preparation of essentially binderless zeolites A using kaolin as a starting material. The kaolin is shaped to the desired form of the finished product as an extrudate or as a sphere. In a subsequent step the formed material is fired at 500–600 °C to convert the kaolin to a metakaolin, and then the agglomerates are treated with hydroxide solution and seeds near the boiling point to convert the kaolin to the desired zeolite. Finally, the agglomerates are washed, dried, and activated to yield the final product [12, 24]. This procedure does not normally lead to a very pure zeolite due to the impurities in the starting material.

$$\underset{\text{Kaolin}}{2\,Al_2Si_2O_5(OH)_4} \longrightarrow \underset{\text{Metakaolin}}{2\,Al_2Si_2O_7} + 4\,H_2O \xrightarrow{NaOH} \underset{\text{Zeolite A}}{Na_{12}[(AlO_2)_{12}(SiO)_{12}] \times 27\,H_2O}$$

A more modern manufacturing method uses crystalline zeolite A powder in the form of a filter cake, which is mixed with the kaolin binder, formed, and calcined. During calcination the kaolin is converted to the more reactive metakaolin, which in turn may be transformed to zeolite A upon treatment with sodium hydroxide. The finished product now consists mainly of pure zeolite A [25, 26]. The pre-formed extrudates or beads may be converted by ion exchange to other forms such as zeolite 5 A. These zeolites have an extraordinarily high adsorption capacity, as they are not diluted with inert binder.

4 Characterization of Zeolites for Adsorption Processes

Production of the whole range of zeolites of a zeolite manufacturing plant is monitored continuously through the production control laboratory. At different stages of the production process samples are withdrawn, analyzed, and characterized in the laboratory. Special equipment is used to obtain all the required information. Part of these instruments and of the techniques used are described in this section. The analyses applied depend very much on the application for which the zeolite is used. For detergent application, zeolite A has to have a uniform particle size, a given morphology, and a high ion exchange capacity, whereas the same type zeolite A has to have a high water adsorption capacity and good hydrothermal stability if it is to be used in a natural gas drying plant. In principle, all methods employed for the investigation of solids can be used for zeolites.

4.1 X-Ray Powder Diffraction

The X-ray diffraction pattern is a typical "fingerprint" which allows one to determine sample purity, degree of crystallinity, or the size of a unit cell of a zeolite. Also, by sampling synthesis mixtures at regular time intervals, it is possible to monitor the progress of a reaction very closely. The determination of the degree of crystallinity is only possible by sample comparison with a standard material. This so-called "crystallinity" is usually determined by comparing the identities of selected reflections [27]. Damage to the zeolite structure by post-synthesis processes like cation exchange, steaming, or calcining can also be assessed using this technique and this equipment. However, two restrictions should be made. As X-ray diffraction is not a very sensitive technique it may not be used to fine-tune a synthesis to produce optimized adsorption processes. For these purposes

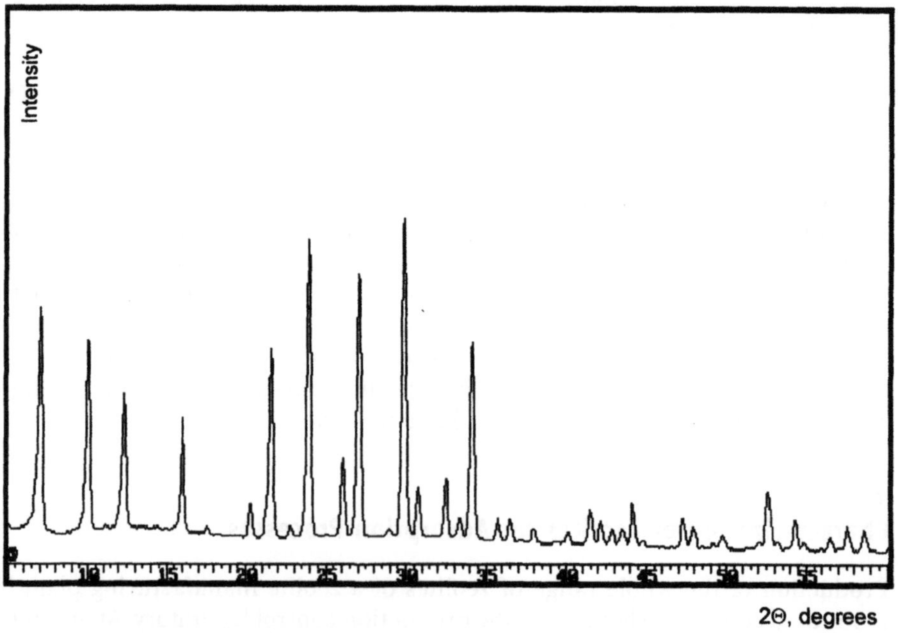

Fig. 10. X-ray diffraction pattern of zeolite A

adsorption measurements are much more reliable and meaningful. After ion exchange, the positions of the peaks in the X-ray pattern do not change, but their intensities do. It is well known that the potassium forms of zeolite A and zeolite X show much lower intensities in their X-ray peaks than their respective sodium counterparts. Figure 10 shows a typical X-ray pattern for zeolite A.

4.2 Elemental Analysis

Elemental analysis is used to determine the chemical composition of a zeolite. Especially interesting are the SiO_2/Al_2O_3 ratio and the composition of the cations present in the zeolite. The SiO_2/Al_2O_3 ratio is important for zeolite A, as it is generally accepted that the stability of this zeolite is higher if this ratio is slightly above 2.00, i.e., between 2.00 and 2.05. Figure 11 shows typical measurements on a number of subsequent batches.

The exchangeable cations – usually alkali or alkaline-earth cations – should exactly counterbalance the amount of aluminum incorporated in the zeolite. Therefore the determination of this number gives a good indication of the quality of the product. Figure 12 gives the typical values of a number of subsequent batches.

There are two different approaches to determine the elemental composition of a zeolite. The first one is to dissolve the zeolite in a suitable matrix and do the subsequent analysis of the solution obtained. It is easy to carry out this wet analysis on powders of the low-silica zeolites A and X, as they are readily soluble in dilute sulfuric acid. The elements silicon, aluminum, and sodium are then determined by classical, well known methods. Atomic emission spectroscopy with inductively coupled plasma (AES-ICP), atomic absorption spectroscopy (AAS), flame photometry, and gravimetric procedures are the methods of choice. However, zeolites formed with a binder or high-silica zeolites do not

Fig. 11. SiO_2/Al_2O_3 ratio of zeolite 4 A

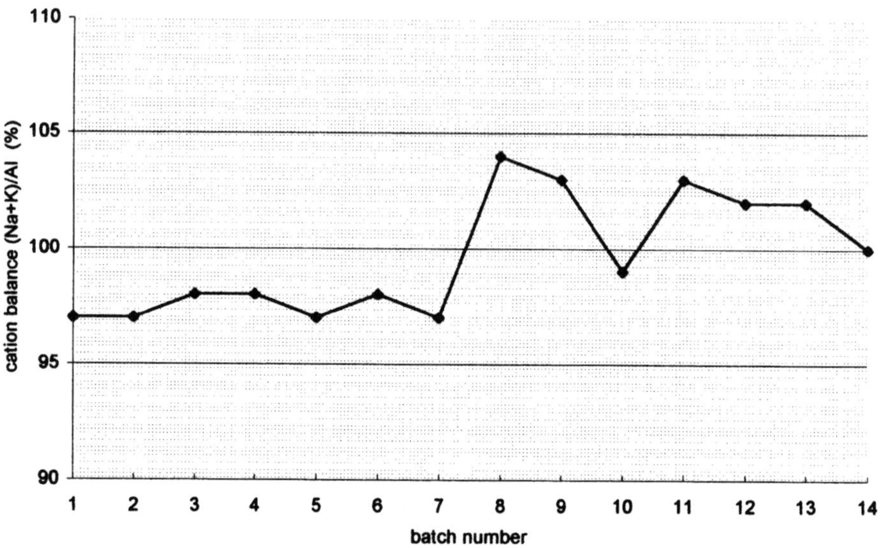

Fig. 12. Cation balance of zeolite 3 A [23]

readily dissolve in dilute acids and must be treated with concentrated hydrofluoric acid to bring them into solution.

Non-destructive physical methods might be an alternative, X-ray fluorescence spectroscopy (XRF) being the most popular. After careful calibration of the instrument it may eventually lead to quantitative measurements. However, the calibration is not easy and only very skilled technicians may produce reliable quantitative results.

4.3 Scanning Electron Microscopy

Scanning electron microscopy (SEM) has become a versatile technique in product development and quality control during zeolite production. With a resolution of a few nanometers, it produces high quality images of single zeolite crystals and of agglomerates. Zeolites with identical chemical compositions and identical degrees of crystallinity may have different geometrical shapes. The morphology is influenced by the conditions used during synthesis. SEM also may show extra-zeolite phases or amorphous compounds. Figure 13 shows a SEM photograph of zeolite 4 A which contains small amounts of sodalite or amorphous material. This extra phase is visible as small spherical particles on the outer surface of the cubic zeolite A crystals. Figure 14 shows a picture of almost perfect, uniformly sized ZSM-5 crystals.

Fig. 13. SEM photograph of Zeolite 4 A

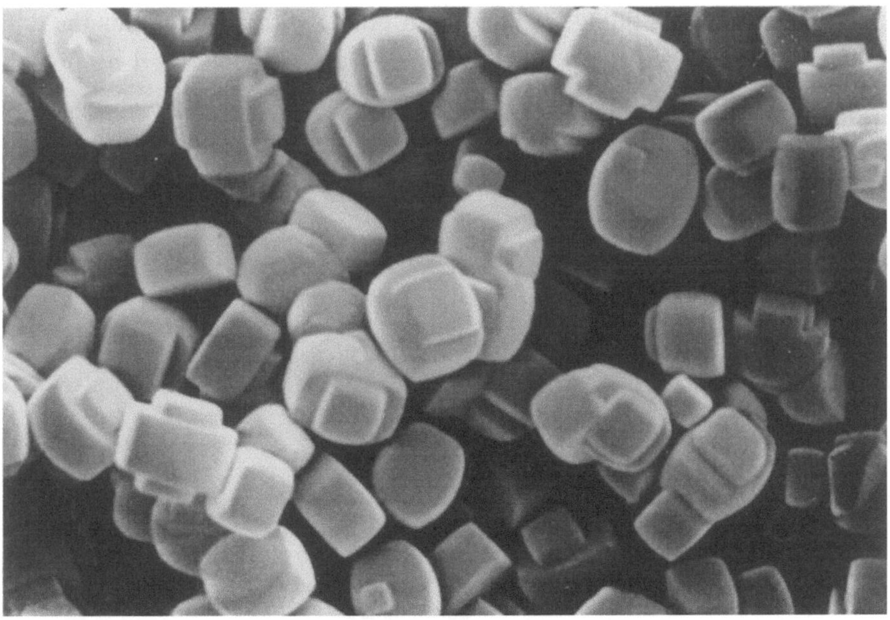

Fig. 14. SEM photograph of uniformly sized zeolite ZSM-5

4.4 Bulk Density

For the engineering of an adsorption plant, it is not only necessary to know the adsorption capacity of the zeolite, but is also important to know the bulk density of the zeolitic material. Such measurements are generally done according to method ISO 787. A measuring cylinder is filled with the formed zeolite, weighed, and then the material is allowed to settle during 1000 cycles of tapping. The volume is determined and the bulk density calculated in the usual way. Figure 15 shows the bulk density of a formed zeolite 3 A that is used for cracked gas drying.

4.5 Dust Formation

A precisely weighed amount of fully activated zeolite beads is put in a cylinder which contains a number of baffles. Air at a defined flow is sucked through the system for 5 min while rotating the cylinder. The air stream removes the dust formed from the rotating cylinder. The dust is transported and deposited on a filter installed between the cylinder and the pump. The filter, which must be dried before use, is weighed before and after the process. The weight increase of the filter is related to the weight of zeolite beads. If the filter is not dried before use it may lose weight during the measurement due to the adsorption of water on the zeolite while blowing air through the system. This would lead to apparently too low figures for dust formation. Full details of the apparatus are given in Fig. 16. Dust formation should be very small for zeolites used in the insulating glass industry. While filling the aluminum profile with zeolite beads the dust

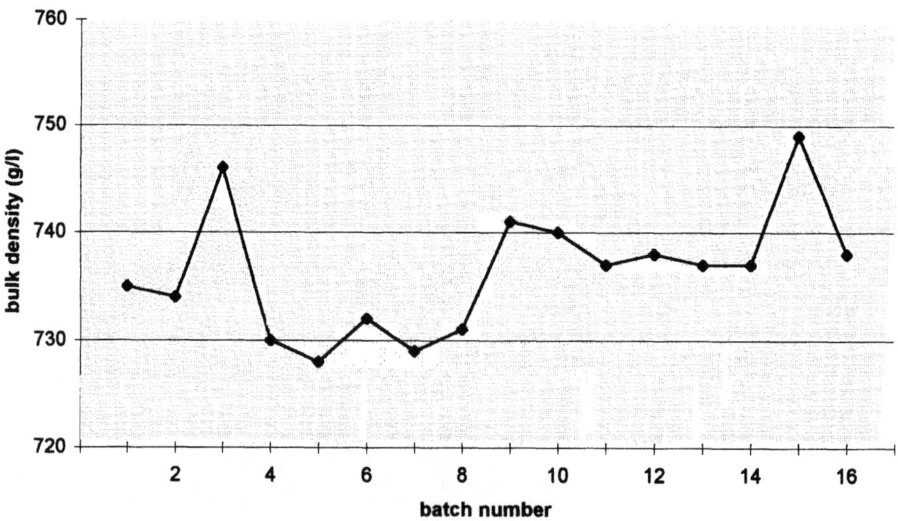

Fig. 15. Bulk density of zeolite 3 A used for cracked gas drying [23]

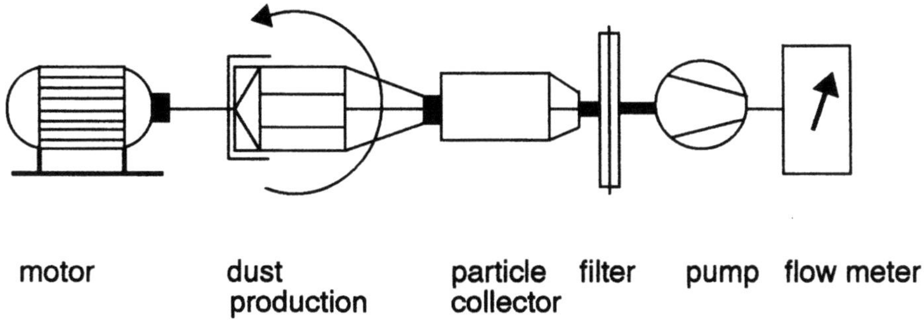

Fig. 16. Equipment for measuring dust formation

might penetrate the small slits and form a deposit on the inner side of the window pane.

4.6 Attrition of Agglomerated Products

A weighed portion of fully activated formed zeolite is placed on a laboratory sieve with a small mesh size. The material is shaken for 15 min in the presence of 5 coins [22]. During the sieving process, the coins will hit the zeolite beads or extrudates and produce a certain amount of fines that are sieved off. The amount of fines formed by attrition gives some indication of the quality of the product. For gas drying units, where the zeolite is used for many cycles of gas drying and regeneration, the attrition value has to be low. In the adsorption tower there is always some friction between the zeolite beads which causes attrition. The formation of fines will increase the pressure drop along the adsorber and the back pressure of the system, which in turn will shorten the lifetime of the adsorbent.

4.7 Bead Size Distribution

For various technical applications bead size and bead size distribution contribute much to the overall performance of the product. In a natural gas drying process, the back pressure becomes higher the smaller the bead size, but on the other hand the mass transfer zone becomes longer the larger the bead size. For this particular application it is a compromise on bead size that leads to an optimized process.

For the application in the insulated glass industry, the maximum bead size is of importance. The aluminum profile is filled automatically through a very small hole. One single zeolite bead of a larger diameter than this hole will block the filling process.

For these reasons, bead size measurement is always an important item in the characterization of zeolites. Bead size measurements may be done through the

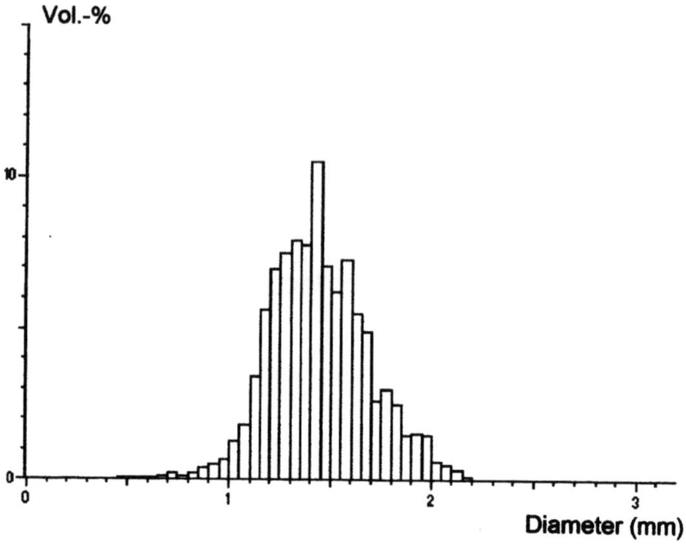

Fig. 17. Typical bead size distribution of zeolite 3 A beads used in the insulating glass industry

same types of sieves as used for the attrition test discussed above. With two sieves of appropriate mesh size, it is possible to determine the undersize and the oversize portion of the bulk material. To get insight into the bead size distribution, more sieves are used to determine the fractions between the undersize and the oversize. In a more modern approach, a video camera and an imaging system are used to determine the bead size distribution and the undersize and oversize of the zeolite beads. This type of analysis gives the required results within a few seconds. A typical bead size distribution is given in Fig. 17.

4.8 Adsorption Characteristics

Adsorption capacity for gases is an important criterion in the application of zeolites. In earlier times, McBain balances were used for the precise determination of adsorption capacity as a function of gas pressure, for both permanent and condensable gases. Samples are placed in small pans inside a glass tube and suspended by a fine quartz spring. After activation of the sample under vacuum, the gas pressure is increased stepwise. The elongation of the spring upon adsorption of gas is proportional to the change in weight. Stepwise measurements at increased gas pressures give the raw data for drafting isotherms for any given combination of zeolite and adsorbed gas at a constant temperature (usually 25 °C). These isotherms yield the fundamental information required for most technical applications of zeolites. The principle of the McBain balance is given in Fig. 18.

An alternative to these measurements are the more modern and more automated volumetric methods. Such types of instruments are commercially avail-

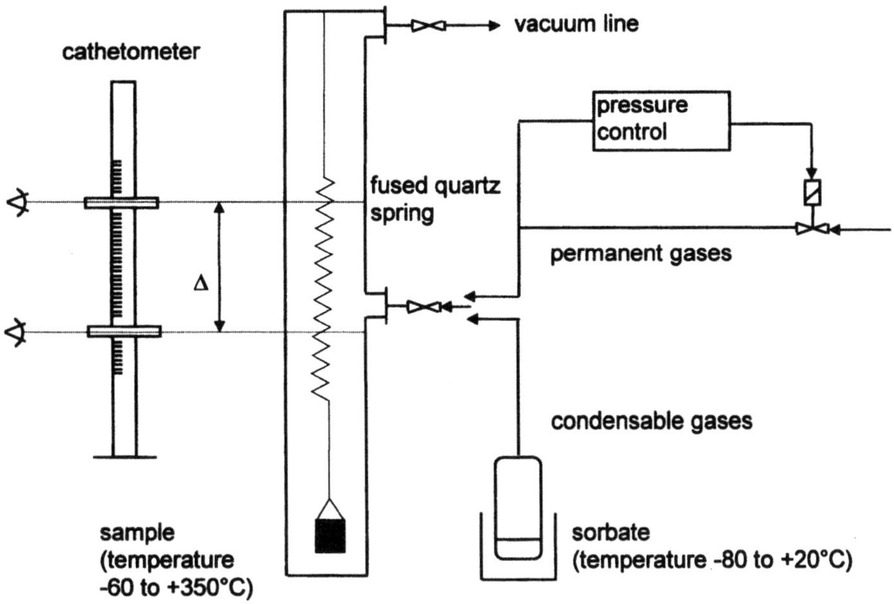

Fig. 18. Function diagram of a McBain balance

able and may produce the whole isotherms. These are fully automated with the support of computers. Instead of measuring the weight, as with the McBain balance, the gas is introduced to the sample in incremental steps, and the pressure difference is measured. Out of these raw data the full isotherm may be calculated.

Thermal gravimetric analyses can provide useful information regarding the thermal treatment of zeolites. Graphs of weight loss vs. temperature often give clear indications as to the temperature required to burn off organic templates or the temperature necessary for a complete or for an optimum regeneration in TSA processes. The highest temperature a zeolite can withstand may also be determined in this way.

4.9 Zeolite Stability

Zeolites have to be stable under operating conditions, meaning that they should not change towards a denser phase nor dealuminate or turn into amorphous compounds. Various methods are available to check this stability, the most versatile being X-ray diffraction and adsorption measurements. In one approach, the X-ray diffraction pattern is determined before and after thermal treatment. Changes in degree of crystallinity, crystal structure changes, or dealumination may be detected. Even more sensitive to changes in zeolite stability are the adsorption values. Using a borderline probe molecule that may penetrate the pores in a pure, undamaged zeolite, the adsorption measurements immediately indicate whether a zeolite was partially changed during thermal treatment. This method, however, does not give any insight into the mechanism of degradation.

In low-silica zeolites, such as zeolite A ($SiO_2/Al_2O_3 = 2.0$) or zeolite X ($SiO_2/Al_2O_3 = 2.5$) structural collapse in a dry atmosphere is observed above 660 °C, while medium-silica zeolites such as zeolite Y ($SiO_2/Al_2O_3 = 5.0$) become amorphous above 700 °C [12]. High-silica zeolites such as ZSM-5 ($SiO_2/Al_2O_3 = 1000$) are structurally stable up to 1000 °C. Above the decomposition temperature, recrystallization to a denser phase usually occurs.

In the presence of water vapor, the decomposition temperature is usually found at much lower temperatures. The hydrothermal stability of a zeolite is of practical importance, as during regeneration the desorbed water of a zeolite builds up a high partial pressure of water. Figure 19 shows the hydrothermal stability of zeolite 4 A and zeolite 3 A [23]. The percent of remaining original water adsorption capacity after steaming at different temperatures is shown.

Figure 20 shows the adsorption capacity for carbon dioxide after having treated zeolite 3 A at different temperatures. A sharp decrease in carbon dioxide adsorption capacity indicates pore closure suitable for cracked gas drying applications [23]. The water adsorption capacity remains unchanged if the time of calcination at higher temperature is chosen carefully.

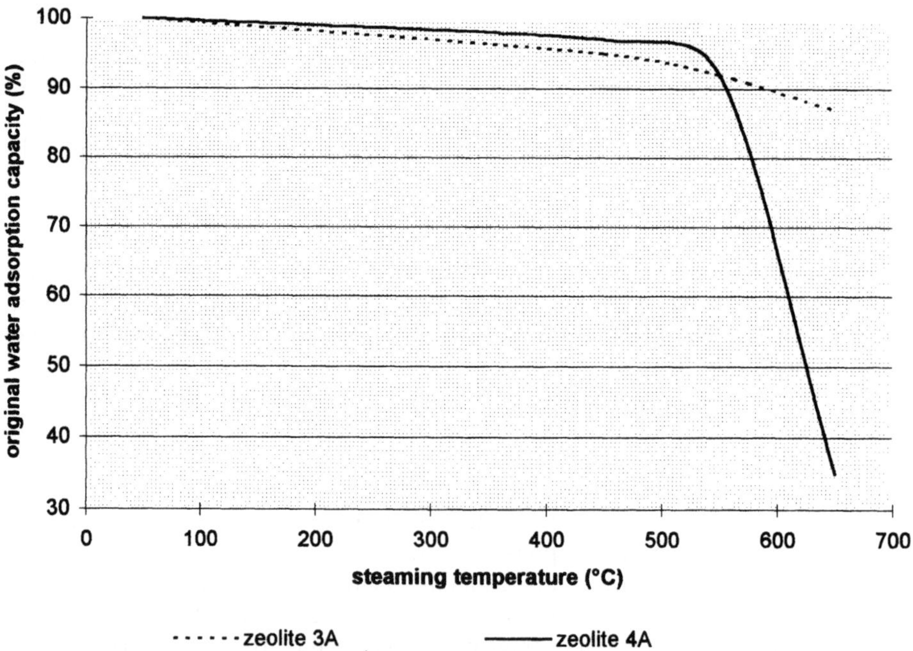

Fig. 19. Hydrothermal stability of zeolite 4 A and zeolite 3 A

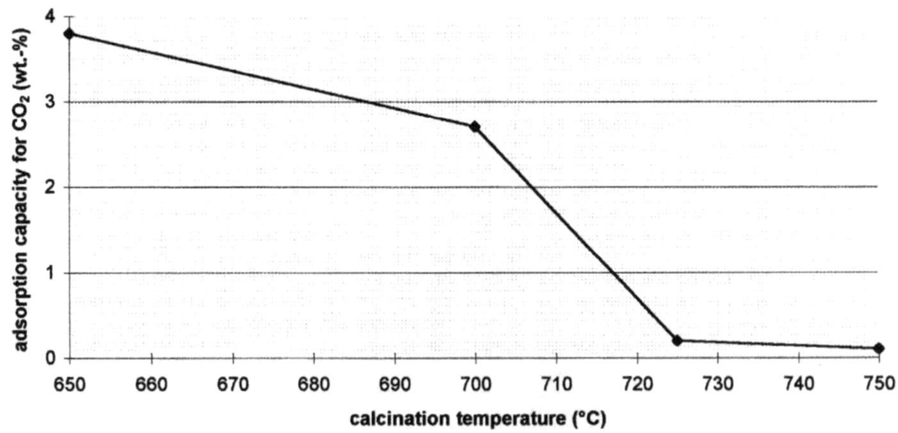

Fig. 20. Influence of the calcination temperature on the adsorption capacity of zeolite 3 A for carbon dioxide

5 Use of Zeolites in Miscellaneous Adsorption Processes

In its powder form, zeolite A is a very effective water scavenger in compounds such as polyurethane systems, epoxy compounds, and metal dust paints. All these compounds will react with water in an unwanted way. Addition of a few percent of zeolite powder to these systems prevents these undesirable reactions, and prolongs the shelf life of these compounds. The sodium form – zeolite 4 A – and the potassium form – zeolite 3 A – are commonly used in these applications. Zeolite 3 A is generally preferred because zeolite 4 A adsorbs nitrogen which is liberated on adsorption of water. This creates undesirable small bubbles within the polymerization mass which impairs product quality. The nitrogen adsorption of zeolite 3 A may be fully suppressed if it is slightly hydrothermally damaged. If this process is carefully done the water adsorption capacity is not negatively influenced.

However, most of the zeolites used in adsorption processes are needed as extrudates or as beads. In such cases, the zeolite is mixed with natural clay binders, wetted with water, formed, and calcined. Beads are usually formed in the size range 0.5–5.0 mm. These beads are applied in a wide range of technical applications, where adsorption of water is one of the most important fields.

5.1 Insulated Glass Industry

Static adsorption in insulating glass plays a major role. Sealed glazing units consist of two (or three) panes of glass separated by a hollow metal spacer. The gas space must be kept free of water vapor and organic vapors to prevent condensation and staining on the inner glass surface. Zeolite beads are inserted in the

Fig. 21. Triple pane window with metal spacer filled with molecular sieve

hollow metal spacer to adsorb all the water present between the window panes (Fig. 21).

The beaded molecular sieves for the insulated glass industry are characterized by four major parameters.

1. The adsorption capacity of the molecular sieve employed should be very high to guarantee a long lifetime of the window pane without condensation of water.
2. The zeolite beads have to be sieved very carefully to the required specifications. The beads are filled into the metal spacer through a relatively small

hole. Oversized molecular sieve beads may plug this hole and interrupt the filling process.
3. The beads have to be free from dust and should not undergo attrition during the filling process. The fines would penetrate the openings of the hollow metal spacer and enter the space between the window panes, leading to an undesired deposit of fines on the glass surface.
4. The molecular sieve should adsorb water exclusively, but no nitrogen, oxygen, or carbon dioxide. If it adsorbed gases in an activated state, it would release such gases while adsorbing water from the glass system. This in turn would build up a pressure between the window panes and eventually lead to glass fracture. For this reason, the major portion of molecular sieve used for this application is of the 3 A type.

A newer version of the insulating glass industry uses foamed silicone rubber profiles as spacers. These profiles are loaded with molecular sieve powder. This system is easier to work with and yields an even better thermal insulation value.

5.2 Drying and Purification of Air and Petrochemicals

Gases and liquids often contain water or other polar contaminants which must be removed before further processing. Natural gas, for example, is usually saturated with water. Before liquefaction it must be dried to a very low dew-point to prevent the formation of water droplets in the plant or during transport. The gas is passed through a fixed bed of activated zeolite beads to remove water down to a dew-point below −70 °C. The saturated zeolite beads are regenerated by heating the bed in a dry gas stream to 230–290 °C. Two or more beds allow continuous operation of such units: one bed is in the adsorption mode, while the other is being regenerated (Fig. 22).

A wide range of liquids and gases are processed using this principle. Purification can include removal of small polar molecules such as carbon dioxide, hydrogen sulfide, alcohols, ketones, mercaptans, and amines. Typical applications of dynamic adsorption processes include:

- drying of petrochemicals and solvents of refineries (ethene, propene, ethylene oxide, natural gas);
- purification of hydrocarbon streams containing sulfur compounds ("sweetening");
- drying and purification of compressed technical gases like air, nitrogen, or rare gases.

The dynamic applications of zeolite X are basically the same as those for zeolite A, except that zeolite X can be used where larger molecules have to be adsorbed. Aromatic or branched aliphatic compounds may enter the pore system of a zeolite X, and therefore this zeolite may be used to separate aromatics from a paraffin stream and for adsorption of mercaptans from natural gas. However, a major use of zeolite X is in the purification of air prior to liquefaction. Zeolite X has a very high adsorption capacity for carbon dioxide, even at low partial pressure.

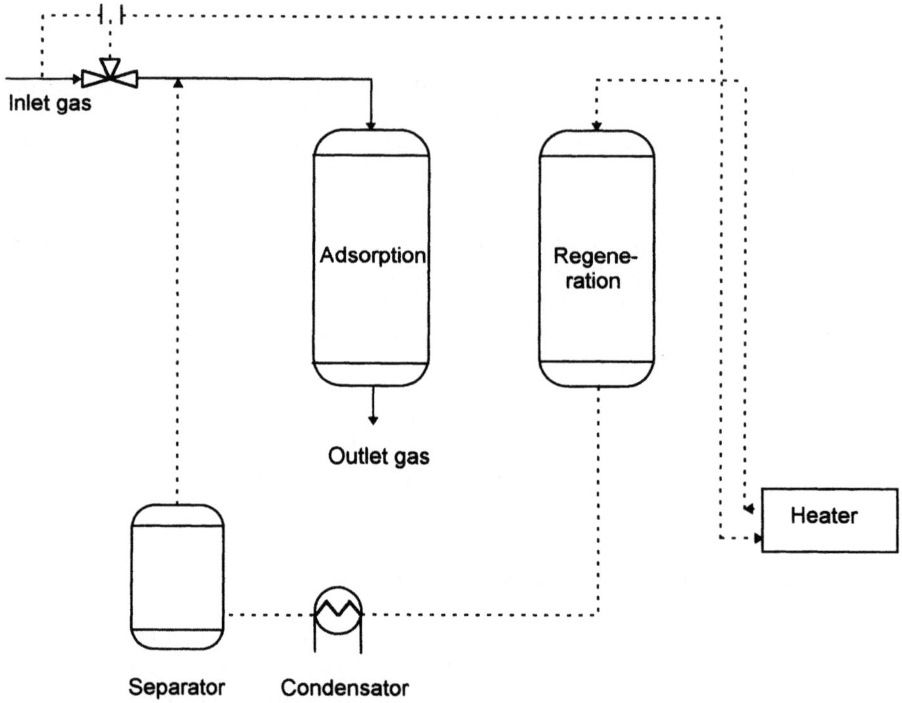

Fig. 22. Dynamic adsorption unit

5.3
Separation of Hydrocarbons

Zeolite 5 A is used in large amounts for the continuous separation of normal and branched alkanes in refineries. The *n*-alkanes may enter the pores of a 5 A zeolite, whereas the branched alkanes cannot (Fig. 23). This process is therefore related to chromatographic processes, and it is performed either in the liquid or

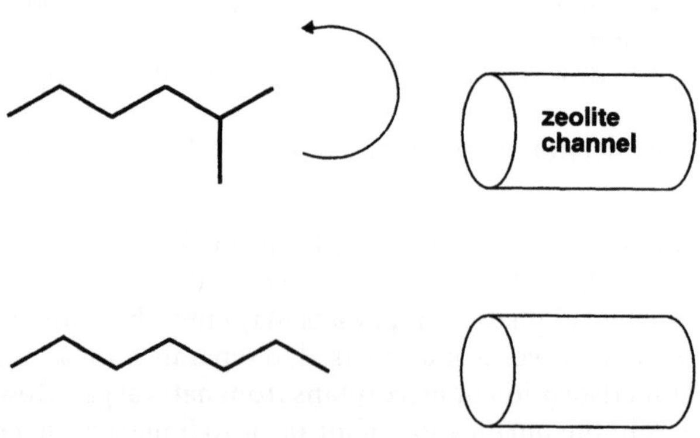

Fig. 23. Separation of *n*-paraffins and iso-paraffins

in the vapor phase at elevated temperatures. Quite often this process is used to upgrade gasoline to a higher octane number. The normal paraffins are then reprocessed in a subsequent process step and mainly used for the production of surfactants. Such processes were developed in the 1950s and were improved in due course. PSA processes in use are developed by UOP (Isosiv), Texaco (TSF), Leuna (Parex-Leuna), ELF (N-Iself), Exxon, and BP [28, 29]. In some of these processes desorption is achieved by displacement with ammonia or short-chain *n*-alkanes. A further development of these processes is the Molex process of UOP, a liquid phase process performed in the countercurrent mode.

5.4 Separation of Gases

Zeolite LiX is especially suited for producing oxygen with a purity of 90–95%. This oxygen quality meets the requirements for oxofuel-fired furnaces, wastewater treatment, ozone production, pulp and paper bleaching, metal cutting, and medical uses. Production costs of 90–93% oxygen produced with a PSA plant are now lower than with cryogenic plants which presently still have a scale-up advantage in larger capacity plants [14]. The vacuum pressure swing adsorption (VPSA) systems can deliver lower-pressure oxygen even more cost-effectively [13]. Oxygen costs are being reduced by continued process improvements.

The separation of oxygen from nitrogen is possible with these materials because the nitrogen molecules, which have a quadrupole moment, are more strongly bound to the zeolite than the oxygen molecules. The purity of oxygen does not exceed 95% as argon cannot be separated with this process.

Acknowledgements. I would like to thank the following colleagues for the help and contributions in various stages of this work: M. Braun, S. Chen, L. Harris, P. Hawes, S. Hitz, B. Kleeb, K. Laube, A. Maggi, J. Strutz, R. Ulrich, and A. Vogt.

References

1. Flanigen EM (1991) Zeolites and molecular sieves – an historical perspective. In: Van Bekkum H, Flanigen EM, Jansen JC (eds) Introduction to zeolite science and practice. Elsevier, Amsterdam, chap 2, p 13
2. Asche W (1996) Chem Rundsch, April 4:1
3. Anonymous (1997) Japan Chemical Week, April 24:4
4. Roland E, Kleinschmit P (1996) Zeolites. In: Ullmann's encyclopedia of industrial chemistry, vol 28 A, 5th edn. VCH, Weinheim, p 475
5. Moscou LM (1991) The zeolite scene. In: Van Bekkum H, Flanigen EM, Jansen JC (eds) Introduction to zeolite science and practice. Elsevier, Amsterdam, p 1
6. Milton RM (1959) US Pat 2,882,244 assigned to Union Carbide Corp
7. Breck DW, Eversole WG, Milton RM, Reed TB, Thomas TL (1956) J Am Chem Soc 78:5963
8. Reed TB, Breck DW (1956) J Am Chem Soc 78:5972
9. Löwenstein W (1954) Am Mineral 39:92
10. Jansen JC (1991) The preparation of molecular sieves. In: Van Bekkum H, Flanigen EM, Jansen JC (eds) Introduction to zeolite science and practice. Elsevier, Amsterdam, chap 4, p 77

11. Barrer RM (1982) Hydrothermal chemistry of zeolites. Academic Press, London
12. Breck DW (1974) Zeolite molecular sieves. Wiley, New York, p 725
13. Monereau C (1996) Sci Tech Froid 147
14. Michael KP (1997) Chem Eng 104:72
15. Chao CC (1989) US Pat 4,859,217 Cl B01D 53/04 assigned to UOP
16. Lechert H, Kacirek H (1991) Zeolites 11:720
17. Kühl GH (1987) Zeolites 7:451
18. Kirner JF (1993) US Pat 5,268,023 Cl B01D 53/04 assigned to Air Products and Chemicals
19. Argauer RJ, Landolt GR (1972) US Pat 3,702,886 Cl C01B 33/28 assigned to Mobil Oil
20. Jacobs PA, Martens JA (1987) Synthesis of high-silica aluminosilicate zeolites. Elsevier, Amsterdam
21. Ribaud GL (1965) US Pat 3,219,590 assigned to Union Carbide
22. Mitchell WJ, Moore WF (1961) US Pat 2,973,327 assigned to Union Carbide
23. Rode EJ, Trent RE, Harris L, Jaussaud D (1997) Molecular sieve adsorbent properties and their impact on cracked gas drying; lecture given at the AIChE Spring Meeting 1997, Houston, Texas
24. Haden WL Jr, Dzierzanowski FJ (1961) US Pat 2,992,068 assigned to Minerals and Chemicals Philipp
25. Chi CW, Hoffman GH, Eichhorn ET (1977) DE Pat 2,707,313 Cl C01B 33/26 assigned to WR Grace
26. Goytisolo JA, Chi DD, Lee H (1975) US Pat 3,906,076 Cl C01B 33/28 assigned to WR Grace
27. Annual book of ASTM standards (1997). Standard test method for determination of relative X-ray diffraction intensities of faujasite-type zeolite-containing materials. American Society for Testing and Materials. West Conshohocken, D 3906-97
28. Ruthven DM (1984) Principles of adsorption and adsorption processes. Wiley, New York, p 396
29. Knoll H, Harms U, Höse W, Fürtig H (1990) Chem Tech (Leipzig) 42:149

Subject Index

A
ABW 9
- framework of 9
adsorbate 18
- localization 18
- *m*-xylene 18
- powder diffraction of 18
adsorbed hydrocarbon molecules 47
- disorder of 47
- location of 47
adsorption of benzene 49
- neutron diffraction data of 49
adsorption processes 163
- zeolites for 163
adsorption site 22
- symmetry of 22
adsorption of pyridine 49
- in gallozeolite-L 49
- in Na-Y 49
- neutron diffraction 49
AES-ICP 185
agglomerated products 189
- attrition 189
air 177
- PSA separation of 177
AlO_4 tetrahedra 126
- isolated 126
$AlPO_4$-5 44, 45
- Al and P ordering 45
- calcination of 44
- distortion to an orthorhombic structure 44
- neutron diffraction by 44
- structure of 44
- X-ray powder diffraction 44
$AlPO_4$-8 (AET) 7, 43
- reconstructive phase transition of 7
$AlPO_4$-11 43
- structure of 43
$AlPO_4$-16 (AST) 17
$AlPO_4$-18 (AEI) 18
- template ordering 18
$AlPO_4$-34 (CHA) 18, 25
- location of the morpholinium cation 25
- morpholinium template 18
- synchrotron diffraction data of 25
- template ordering 18
$AlPO_4$-54 43
- structure refined from neutron powder data 43
- structure of 43
- water molecules in 43
$AlPO_4$H2, (AHT) 7
- octahedral coordination of aluminum 7
- phase transition to tridymite 7
AlPOs 42, 43, 142
- neutron powder diffraction by 43
- ordering in 43
- overlapping reflections 43
- peak broadening 43
- scattering lengths 43
AlPO-structures 148
- even-membered rings in 148
- local charge balance in 148
- ordered T-atoms in 148
alumina sources 170, 179
aluminate sodalite 128
aluminate sodalites 119, 131
- structure of 131
- tetrahedron distortion in 119
aluminogallosilicate, $LiAl_{0.5}Ga_{0.5}SiO_4 \cdot H_2O$ 39
- disordered Ga/Al distribution 39
- positions of the water molecule 39
aluminogermanate sodalite 120
aluminophosphates 42
aluminosilicate 42
- neutron powder diffraction by 42
aluminum hydrate 170
1-aminoadamantane 17
- spherical electron density 17
analcime 39, 57, 144
- leucite 144

analcime
- neutron diffraction by 57
- neutron single-crystal study of 39
- pollucite 144
- wairakite 144
angular distortion 126
aniline 53
- adsorption sites by molecular mechanics calculations 53
- in Yb, Na-Y 53
antiphase domains 130
apertures 147, 157
- free diameters of 147, 157
- shape of 147
Ar 57
- adsorbed into $AlPO_4$-5 57
- adsorbed into $SAPO_4$-5 57
- neutron diffraction by 57
argon 56
- in AFI 56
- in MFI 56
- in silicalite I 56
aristotype 114, 125
atlas of zeolite structure types 142
atomic absorption spectroscopy (AAS) 185
atomic emission spectroscopy 185
- with inductively coupled plasma
atomic scattering factors 73
attapulgite 183
attapulgite-type clays 181
attrition 189, 195
- of agglomerated products 189
attrition resistance 180

B
Ba-X zeolite 52
- higher selectivity for *p*-xylene 52
barrerite 60
- framework structure of 60
- neutron diffraction of 60
BEA 99, 100
- ED pattern of 99
- framework structures of 100
- HRTEM images of 99
- polymorphs of 99
- stacking disorder of 99
- structure of 99
- structural faulting of 99
bead size distribution 189
beads 195
- oversized 195
bentonite 181, 183
benzene 47, 48, 53–55
- adsorption sites by molecular mechanics calculations 53
- by powder diffraction 47
- clusters of 48
- diffusion of 49
- disorder of 55
- in dehydrated sodium zeolite Y 47
- in faujasite structures 47
- in H-SAPO-37 48
- in H-Y 49
- in MFI structure 55
- in Na-Y 48
- in potassium zeolite L 54
- in ZSM-5 55, 56
- location of 55, 56
- molecular motion of 55
- neutron diffraction of 55
- neutron scattering of 48
- NMR investigations of 48
- orientations 55
- spectroscopic measurements of 48
- X-ray diffraction of 55
berylloarsenate 42
- neutron powder diffraction of 42
beryllophosphate 42
- neutron powder diffraction of 42
beryllophosphate, $Mg_{19}Na_{58}(BePO_4)_{96}$ 46
- alternation of Be and P in 46
- faujasite structure of 46
beryllophosphate-H, BPH 6
beryllosilicate 150
- 3-rings in 150
beta 23
- accessibility 23
- faulting of 23
- polymorphs of 23
- powder pattern of 23
- structure of 23
bikitaite, $Li_2Al_2Si_4O_{12} \cdot 2\,H_2O$ 62
- chains of hydrogen-bonded water molecules in 62
- neutron diffraction of 62
- proton conduction in 63
- Si/Al ordering in 62
- structure of 62
- water molecules in 62
binders for zeolites 183
- characteristic data of 183
block-waves 130
bond anharmonicity 127
bond length distortion 122, 126
bond lengths and angles 116
borosilicate RUB-13 (RTH) 24
- end members of 24
- protonated water 65

brewsterite, $(Sr,Ba)_2[Al_4Si_{12}O_{32}] \cdot 10\ H_2O$ 60
- neutron diffraction of 60
- water molecules in 60
Brillouin zone 134–136
Brønsted acid sites 63
Brønsted sites in zeolites 65
bulk density 188

C
carbon monoxide 57
- in MFI 57
cation positions 23
- artifact caused by Fourier technique 23
- in Linde Q 23
- in sodalites 23
cations 22
- positions in the structure 22
- spherical symmetry of 22
CF_4 57
- adsorbed into $AlPO_4$-5 57
- adsorbed into $SAPO_4$-5 57
- neutron diffraction of 57
CH_4 57
- adsorbed into $AlPO_4$-5 57
- adsorbed into $SAPO_4$-5 57
- neutron diffraction of 57
channels 147
- free diameters of 147
chaotic structure 130, 134, 136
chemical information 66
- bond distance and angle restraints 66
classification systems 151, 154, 155
- criteria 154
- smallest ring(s) 154
- structure types 154
clathrasil-type frameworks 152
clays 181
- for binding zeolites 181
clinoptilotite 60, 164, 166
- neutron diffraction of 60
- thermal stability of 60
cloverite (-CLO) 85, 86
- HRTEM of 86
- low and high magnification images of 86
- scanning EM images of 85
clusters 77, 94
- HRTEM image of 94
- location in zeolites 94
coadsorption in Na-Y 54
- aniline 54
- *m*-dinitrobenzene 54
CoAPO-5 (AFI) 25
- isomorphous substitution 25
- X-ray diffraction data of 25
CoAPO-50 (AFY) 6
CoGaPO-5 (CGF) 18
- template ordering 18
coherent scattering 34
- interference effects 34
complementarity 66
- of X-ray and neutron data 66
configuration 142
- apertures 142
- cages 142
- chains 142
- loop configurations 142
- secondary building units 142
conformations 157, 158
- changes 158
- connectivity 158
- of a ring 158
conformational change 159
- diffraction pattern 159
conformations of 142
connectivity 142
contrast 81
contrast transfer function 81, 82
- imaginary part of 82
coordination sequences (CS) 156, 157
- application of 157
- TD-values 157
CoSAPO-34 (CHA) 25
- isomorphous substitution in 25
- X-ray diffraction data of 25
crush strength 181
crystal growth units 92
crystal size 25, 172
- limitations of 25
crystal structures 1, 2, 36
- accurate refinement 36
- determination of 2
- by X-ray diffraction 1
crystallographic problems 2
- crystal size 2
- disorder 2
- faulting 2
- framework flexibility 2
- isomorphous replacement 2
- overlap of reflections 2
- space group 2
CSI 155

D
D-RHO 42, 64
- framework of 42
- flexibility of 42
- hydroxyl groups of 64
- position of the deuterium atom in 64

D-Y 64
- deuterium atoms in 64
- neutron powder diffraction of 64
DAC 159
- prediction of 159
dark field imaging 75
dealumination 98
- of the framework 98
- studied by HRTEM 98
Debye-Waller factor 36
decasil RUB-3 (RTE) 24
- end members of 24
diffracted beams 81
- phase of 81
deuterium 56
- in zeolite 13X 56
diffraction behavior 73, 74
- electrons 73, 74
- neutrons 73, 74
- X-rays 73, 74
diffraction data 11
- symmetry changes 11
diffraction pattern 36, 74, 79
- Fourier transform of 79
- powder case 36
- single crystal case 36
diffuse streaks 134
dislocations 115
displacive distortion waves 130
distance least squares (DLS) refinement 6, 160
- theoretical networks 160
dodecasil 3C 134
dodecasil 3C-tetrahydrofuran 126
dodecasil-1H (DOH) 17, 25
- population factors of 25
DOR-NMR 120
drying and purification 195
dust formation 188

E
ECR-30 155
ED patterns 79, 109
- of ERI 79
- of LTL 79
- of OFF 79
- of TPA/ZSM-5 109
- quantitative analysis of 109
edingtonite, $Ba_2Al_4Si_6O_{20} \cdot 7\,H_2O$ 58
- neutron diffraction of 58
- phase transition of 58
- Si/Al ordering of 58
- water molecules in 58
electron atomic scattering factors 107
- of light elements 107
electron charge distribution 73
electron crystallography 75, 106
- electron atomic scattering factors 106
electron diffraction (ED) 3, 73, 74
- intensity 74
electron microscopy 77–79, 104
- of BEA 77
- of clusters 77
- of crystal growth 77
- of crystallinity of zeolites 77
- of defects 77
- electron diffraction patterns 104
- of EMT/FAU 77
- of ERI/OFF 77
- of ETS-10 77
- of fine structure of zeolites 79
- of LTL 77
- of intergrowths 77
- of mesoporous materials 104
- of natural mineral FAU 78
- of polymorphs 77
- of surface structures of zeolites 77
electron microscopy on zeolites 78
- historical view 78
electron radiation damage 104
- knock-on damage 104
- radiolytic damage 104
electron wave field 79
- Fourier transform of 79
electrons 73, 74, 80, 81
- aberration 73
- atomic scattering factor for 73
- crystal structure factor for 81
- diffraction behavior 74
- image 74
- lenses for 73
- Lorentz force on 73
- in magnetic fields 73
- phase shift 80
- refractive index 80
- wavelength of 73, 82
elemental analysis 185
ellipticity parameter (EL) 10
EM 77, 82, 84
- defects 77
- resolution of 82
- TV system for 84
EM image 73
EMC-2 (EMT) 18
- template ordering in 18
EMT 24, 88, 90, 98, 99, 155, 159
- faujasite sheet 88
- HRTEM 90
- HRTEM images after dealumination 98, 99

- intergrowth of 88, 90
- prediction of 159
- stacking faults in 155

EMT/FAU 24
- end members of 24

ERI/OFF 24
- end members of 24

ETS-10 99-104
- chiral crystal system 104
- ED patterns 100, 101
- electron microscopy of 99
- faults of 100
- framework composition ratio of Si/Ti = 5 102
- framework structures of 100
- high resolution TEM (HRTEM) images of 99, 100-102
- models for the structures of 103
- polymorphs of 99
- SEM image of 100
- stacking sequences of 102
- structure of 99
- structure unit of 102

extra-framework sites 23
extrudates 182

F

fast iterative patterson squaring (FIPS) 4
FAU 24, 85, 86, 88, 90, 92-94, 96, 98, 99, 104, 155
- clusters of Fe_2O_3 in 96
- clusters of MoO_3 in 96
- clusters of MoS_2 in 96
- crystal growth in 94
- D6R unit of 94
- electron damage of 104
- end members of 24
- faujasite sheet 88
- growing steps of 92
- high resolution TEM surface profile images of 93
- HRTEM images 90, 92
- HRTEM images after dealumination 98, 99
- intergrowth of 88, 90
- metal compounds in 96
- scanning EM images of 85
- Si/Al ordering in 94
- simulated images of 93
- surface model of 93
- surface steps on 86
- surface structure of 92
- surface structure models of 94
- stacking faults of 155
- twins in 86

FAU/EMT 23
- end members of 24
- faulting in 23

faujasite sheet 89, 92
- crystal growth mechanism 92
- EMT 89
- FAU 89

fault planes (FP) 151, 152, 155
- table 155

faulted intergrowths 24
- blocking of the pores 24

faulting 23
- irregular 23
- new framework topology 23
- powder pattern 23

faulting planes (FP) 160
- theoretical networks 160

feldspars 123, 130
- modulations occurring in 130

ferrierite 42
- energy minimization of 42
- neutron powder diffraction of 42
- Si MAS NMR of 42

ferroelastic strain 130
ferroelectric polarization 130
fertilizers 166
- additives for 166

fibrous zeolites 124
filler 166
- in the paper industry 166

fine structures 87
- as-synthesized LTL 87
- of zeolites 77

flame photometry 185
flexibility 10
- of NAT 10

fluid catalytic cracking (FCC) 165
fluoride 170
- as mineralizing agent 170

forming processes 180
four-connected nets 115
Fourier recycling procedure 5
framework 6, 7, 158
- density of 146
- distortions of 6, 7
- rigidity of 158

framework atoms 24
- isomorphous replacement 24

framework conformation 157
framework densities (FD) 146, 148, 149
- criterion of 148
- normalized FD-values 148
- smallest rings 149
- table 148

framework flexibility 8
- in response to different cations 8
- in response to non-framework species 8
framework structures 1, 31, 159
- of new zeolites 31
- prediction of 159
- theoretical 159
- X-ray diffraction of 1, 31
framework topologies 153, 154
- classification system of 154
framework topology 6
- space group 6
frameworks 8, 122, 123, 148, 150, 160
- of ABS 8
- collapsible 8, 123
- conformational changes of 122
- dense 148
- finite units of 150
- of FAU 8
- of GIS 8
- hypothetical 160
- of KFI 8
- of LTA 8
- of $MAlSiO_4$ 8
- of MEL 8
- of MFI 8
- of NAT 8
- non-collapsible 8, 123
- porous 148
- of RHO 8
- of SOD 8
- of TOT 8
- table of 150
Fraunhofer diffraction pattern 79
FSM 104
- electron microscopy of 104
- electron diffraction patterns of 104
FSM-16 104, 107–109
- calcination of 109
- growth mechanism 109
- HRTEM of 104
- HRTEM images of 107
- length along the channels of 109
- morphology of 107
- observed HRTEM image of 108
- shape of channels in 107
- simulated image of 108
- wall thickness of 107
- X-ray diffraction of 109

G
gallophosphite 66
- structure of 66
gallophosphite, Ga_2-$(HPO_3)_3 \cdot H_2O$ 46
- hydrogen-atom position in 46
- neutron data of 46
- octahedrally coordinated Ga atom in 46
- structure of 46
- X-ray data of 46
gallosilicates 38, 39
- occupancies of various sodium-ion sites in 39
- structures of 39
- with the FAU framework 39
$GaPO_4(OH)_{0.25}$ (-CLO) 5
- large single crystal of 5
- structure determination of 5
gases 195, 197
- drying and purification of 195
- separation of 197
gel 172
- aging of 172
GIS 8, 158, 159
- conformations of 158
- framework of 8
- XRD powder patterns of 159
gismondine 9, 62, 63, 159
- cation coordination in 61
- framework deformation of 9
- hydrogen bonding in 61
- water molecules of 61
- XRD powder patterns of 159
gismondine, $Ca_{3.91}Al_{7.77}Si_{8.22}O_{32} \cdot 17.6H_2O$ 60
- neutron diffraction by 60
- ordered Si/Al distribution in 60
- X-ray refinement of 60
gismondine-type framework 8
GME 155
- stacking faults in 155
granulation wheel 180
group–maximal subgroup relationships 119
guest molecules 16, 17
- spherical electron density of 17
- symmetry 16

H
H_2 57
- adsorbed into $AlPO_4$-5 57
- adsorbed into $SAPO_4$-5 57
- neutron diffraction 57
H-SAPO-34 65
- neutron diffraction of 65
- existence of hydronium ions in 65
H-SAPO-37 49, 64
- Al/P ordering in 49
- benzene adsorption in 64
- protons in 64

H-Y 64
- hydrogen positions in 64
H-ZSM-5 (MFI) 11, 13, 14, 18-21, 65, 187
- adsorbed molecules in 21
- application of mechanical stress 11
- ferroelastic 13
- incoherent inelastic neutron scattering by 65
- loaded with naphthalene 20
- mechanical stress in 14
- monoclinic/orthorhombic symmetry change of 14
- monoclinic symmetry of 11
- naphthalene in 19, 20
- neutral complex of water in 65
- organic molecules in 18
- orientation of adsorbate in 20
- orthorhombic/monoclinic transition of 13
- orthorhombic symmetry of 11
- *p*-dichlorobenzene in 18-21
- *p*-nitroaniline in 20
- *p*-xylene in 20, 21
- pentasil layer 11
- phase transition of 11
- rotational orientations of adsorbed molecules in 21
- SEM photograph of 187
- sorbate free 14
- sorbate loaded 14
- twin domains in 11, 13
H-ZSM-5 crystal 13
- rotation photographs of 13
hauyne 124, 134
hettotypes 114
heulandite 164
heulandite $Ca_4Al_8Si_{28}O_{72} \cdot 24\ H_2O$ 58
- neutron diffraction by 58
high resolution SEM 85
high resolution TEM 87, 89
- of EMT 89
- of FAU 89
high-resolution transmission electron microscope (HRTEM) images 75
high-voltage HRTEM 106
Hoffmann degradation 170
HREM imaging 80
- schematic diagram of 80
HRTEM 79, 87, 94
- of LTL 87
- of serial ultra thin sectioned specimens 94
- of particles 94
- of zeolites 79
HRTEM images 75, 78, 81, 82, 84, 88-90, 92, 95-99, 107, 109
- best focus conditions 84
- of channels 89
- of clusters 75
- contrast of 82
- correcting astigmatism 84
- of dealuminated EMT 99
- of dealuminated FAU 98
- of EMT 98
- of EMT/FAU 90
- of FAU 92, 98
- fine structure by 78
- of FSM-16 107
- impulse response function 82
- intensity distribution in 81
- of intergrowth structures 89
- of LTL 88
- of MCM-41 107
- of mesopores 98
- of mesoporous materials 107
- of modulation of Se-content 97
- of MoS_x/FAU 96
- nature of disorder by 78
- potential of the crystal 82
- of Pt/LTL 95
- quantitative analysis of 109
- of Se-MOR 97
HRTEM recording 84
- lower magnification 84
hydrocarbons 196
- separation of 196
hydrocracking 165
hydrogen 56
- in zeolite 13X 56
hydrogen bond acceptors 63
hydrogen bonding 38
hydronium ion 64
hydronium species 63, 65
hydrothermal stability 192
hydroxide 170
- as mineralizing agent 170

I
ideal and real structures 114
image by electrons 74
image processing 94, 106
- artificial contrast 106
- of particles 94
incoherent background 35
incoherent scattering 34
- dynamics of 34
indexing programs 3
inelastic neutron spectroscopy 34

inelastic scattering 75
- of electrons 75
insulated glass industry 193
insulation glass 165
- zeolite A in 165
intensities 4
- partitioning 4
intergrowth 89
- of EMT 89
- of FAU 89
- oscillatory growth in 89
intermediates 155
- of FAU/EMT 155
- of Linde T (ERI/OFF) 155
interrupted frameworks 142, 144
- 3-letter code 144
- missing T–O–T bridges 142
ion exchange 177
ionic conductivities 40
- in X 40
- in Y 40
isomorphous replacement 24
- change in unit cell volume by 24
- distribution of the incorporated tetrahedral atoms 24
- of framework atoms 24
isomorphous substitution 24
- population parameter 24
Isosiv process 197
isotropic temperature factor 36
isotropy subgroups 118
isotypes 144, 146
- list of 146
isotypic pairs 148
- of phosphates 148
- of silicates 148
IZA Structure Commission 142

K
kaolin 183
kaolin binder 177
kaolin-type clays 181
krypton 56
- in AFI 56
- in MFI 56
- in silicalite I 56

L
La-Y 64
- hydroxyl group attached to the La^{3+} 64
- proton positions in 64
large-pore zeolites 134
- modulated phases in 134
laumontite, $Ca_2Al_4Si_8O_{24} \cdot 16\, H_2O$ 63
- dehydration/rehydration of 63
- disorder of water in 63
- neutron diffraction by 63
- ordered Si/Al in 63
- water molecules in 63
lazurite 134
leonhardite 57
- neutron diffraction by 57
$LiBeAsO_4 \cdot H_2O$ 8
$LiBePO_4 \cdot H_2O$ 8
$LiGaAlSiO_4 \cdot H_2O$ 8
$LiGaSiO_4 \cdot H_2O$ 38, 46
- bond lengths in 38
- framework ordering in 38
- hydrogen bonding in 38
- coordination of lithium in 38
- positions of the hydrogen atom in 38
- water in 38
lignosulfonate 182
lithium exchanged zeolite X 177
$LiZnAsO_4 \cdot H_2O$ 8
$LiZnPO_4 \cdot H_2O$ 44
- neutron diffraction by 44
$LiZnPO_4 \cdot H_2O$ 8
localization of particles by HRTEM 94
lock-in phase transitions 129
longitudinal strain 120
loop configurations 116, 155–157, 160
- immediate surrounding of the T-atoms 156
- theoretical networks 160
lovdarite (LOV) 24, 150, 152
- end members of 24
- 3-rings 150
Löwenstein's rule 168, 176
LTA 15, 85, 86
- FAU crystals overgrowth 86
- possible distortions in 16
- scanning EM images of 85
- SEM image of 86
- topology in 15
LTA framework 121
- idealized 121
LTL 87, 88, 95, 105
- anisotropic damage under an electron beam 105
- boundaries in 88
- channels observed in 87
- faults in 88
- HRTEM image of 87
- Moiré effects in 88
- blockage of one-dimensional channels 88
- Pt-containing 95
- radiation damage process of 105

M
m-dinitrobenzene 53, 54
- adsorption sites by molecular mechanics calculations 53
- in Na-Y 54
- neutron diffraction of 54
m-nitroaniline 53, 54
- adsorption sites by molecular mechanics calculations 53
- in Na-Y 54
m-xylene 51, 53
- in Ba-X 53
- in Na, Yb-Y 51
- quasielastic neutron scattering (QENS) of 51
- rotational dynamics 51
$M_{4-x}HGe_7O_{16} \cdot nH_2O$ 46
- neutron diffraction by 46
- octahedra in the framework of 46
- powder X-ray data of 46
- structure of 46
magnetic scattering 32
- ordering of unpaired electron spin density 32
MAPSO-46 (AFS) 6
McBain balance 191
MCM 104
- electron microscopy of 104
- electron diffraction patterns of 104
MCM-22 107
- electron crystallography of 107
- structure of 107
MCM-41 104
- electron microscopy of 104
- HRTEM of 104
MCM-48 104
- electron microscopy of 104
mechanical stress 14
MEL 24, 89, 91-93, 155
- ED pattern of 91, 93
- framework structure of 91
- HRTEM image of 91, 93
- post synthetically dealuminated 89
- projections of the framework structure of 92
- simulated ED patterns of 92
- stacking faults of 155
- structures of 89
MEL/MFI 24
- end members of 24
membrane zeolites 106
- high-voltage HRTEM of 106
MEP 152
mesitylene (1,3,5-trimethylbenzene) 53
- adsorption sites by molecular mechanics calculations 53
- in Yb, Na-Y 53
metakaolin 183
methane 56
- in AFI 56
- in MFI 56
- adsorption sites by molecular mechanics calculations 53
MFI 24, 89, 91, 155
- end members of 24
- projections of the framework structure 91
- simulated ED patterns of 91
- stacking faults in 155
- structures of 89
MFI/MEL 23
- faulting in 23
$(Mg,Al)PO_4$-STA-1 (SAO) 25
- synchrotron diffraction data of 25
mineralizing agent 168
modulations of structures 113, 128, 130
- cause for 130
- commensurate 128
- incommensurate 128
modulation vector 129
modulation wave 128
- in microporous materials 113
- wavelength of 128
moiré effects 88
molecular sieves 71
- electron microscopy of 71
Molex process 197
montmorillonite 182
mordenite (MOR) 97, 166
- acid-washed 166
- Se-chains in 97

N
N-Iself process 197
Na, H-Y 64, 65
- hydrogen atoms positions in 64
- neutron diffraction of 64
- proton positions in 65
Na-A zeolite 127
- dehydrated 127
NaP1 158, 159
- conformations of 158
- isotype of GIS 158
- XRD powder patterns of 159
NAT 10
- frameworks collapse in 10
- single crystal diffraction of 10
- topology of 10

natrolite 57, 120
- neutron diffraction of 57
- ordered Si/Al distributions in 57
- water in 57
natrolite, $Na_2Al_2Si_3O_{10} \cdot 2\,H_2O$ 57
- water molecules in 57
- X-ray diffraction of 57
natural gas 167
- drying of 167
natural hydrated minerals 57
- hydrogen bonding in 57
- neutron diffraction of 57
- Si/Al order or disorder in 57
- single-crystal studies of 57
- water molecules in 57
natural zeolite market 166
natural zeolites 79, 166
- as dietary supplements 166
- scanning electron microscopy of 79
neutron data 44
- accuracy of structural refinements 44
neutron diffraction 31, 32, 38, 40, 66
- arrangements of water molecules 32
- cations 38
- framework 38
- cations in the cavities 32
- hydrogen bonds 32
- information by 66
- by Li-X 40
- by Li-Y 40
- location of adsorbed molecules 32
- ordering of Si and Al 38
- powders 31
- refinements 66
- single crystals 31
- structural information 31
neutron scattering 33
- elastic 33
neutron scattering power 33
neutrons 32, 35, 36, 74
- diffraction behavior 74
- flux 35
- long-wavelength 36
- monochromatic beam of 35, 36
- pulsed source 35
- short-wavelength 36
- source 35
- steady-state source 35
- wavelength of 36
nitrogen 57
- in MFI 57
nonasil (NON) 18
- cobalticinium template ordering 18
non-framework species 16
- adsorbed molecules 16
- cations 16
- disorder of 16
- localization of 16
- partial occupancies 16
- pseudo-symmetry of 16
- templates 16
non-templated synthesis 180
normal- and iso-paraffin separation 167
nosean 134
Nu-1-type frameworks (RUT) 24
- isomorphous replacement in 24
nuclear neutron diffraction 35
- characterization of disordered systems by 35
- differentiation between neighboring elements by 35
- localization of light atoms by 35
nuclear scattering 32
- arrangement of the atomic nuclei 32
nuclear scattering length 33

O

objective aperture 81
OFF 24
OFF/ERI 23
- end members of 24
- faulting 23
OTO bond angles 117
overlap of reflections 3

P

p-xylene 51, 52
- hydrogen bond type interactions of 52
- in Ba zeolite X 52
- in Na, Yb-Y 51
- in Na-Y zeolite 52
- molecular simulation 52
- quasielastic neutron scattering (QENS) 51
- rotational dynamics of 51, 52
pahasapaite 10
- powder data of 10
- single crystal structure of 10
pane window 194
- filled with molecular sieve 194
Parex-Leuna process 197
partial collapse 123
- of natrolite and sodalite 123
particle size 94
particles 94
- HRTEM images of 94
- localization by HRTEM 94
partitioning the intensities 4
- direct methods 4
peak overlap 37

peak position of X-ray reflections 3
- determination of 3
peak-shape function 36
perovskites 115, 123
petrochemicals 195
- drying of 195
phase shift 81
- by objective lens 81
phase transitions 11, 20, 119, 128
- ferroelastic 128
- ordering of the sorbed phase 20
- splitting of the reflection spots 11
- structural 119
platinum particles 95
- HRTEM images of 95
point groups 118
pollucite 39
polymerization 134
- of guests in channels 134
pore openings 159
- effect of conformation 159
powder data 2, 5, 36
- accuracy of 2
- overlapping peaks 36
- reflections 5
powder diffraction 66
powder neutron diffraction 32
powder patterns 3, 6
- observed 6
- simulated 6
- structure determination via 3
powder structure determination 8
- cell dimension 8
- symmetry 8
precipitated amorphous silicas 170
pressure swing adsorption (PSA) 165
protons in zeolites 63
- location of 63
- neutron diffraction by 63
- position of 64
- via X-ray-diffraction 63
pseudo-symmetric structures 115
Pt clusters 95
- HRTEM images of 95
Pt particles 96
- HRTEM image of 96
- lattice fringes of 96
pyrogenic silica 169

Q
quadrupole coupling constant 120
quantum materials 77
quasielastic neutron scattering (QENS) 35
- rotational and translational motions 35
quaternary ammonium cations 170
quinuclidine 17
- spherical electron density of 17

R
rare gases 56
- in AFI 56
- in MFI 56
- localized adsorption 56
$Rb_{10}Al_{10}Si_{38}O_{96}$ 41
- location of Rb 41
- Rho framework 41
- water molecules in 41
$Rb_{24}Be_{24}As_{24}O_{96} \cdot 3.2\, D_2O$ 41
- location of Rb 41
- Rho framework of 41
- water molecules in 41
reciprocal lattice vector 129
reflections 3-5, 11
- indexed 3
- overlapping 4, 5, 11
- space group 5
refrigerants 167
- drying of 167
related frameworks 150
- finite units in 150
- table of 150
related structure types 151
related structures (polytypes) 155
resolution 82-84
- aberration 82, 83
- contrast transfer function (CTF) 83
- electron probe size 83
- HRTEM 83
- Rayleigh criterion 83
- reconstruction 83
- SEM 82, 83
- TEM 82
- for the study of fine structures of zeolite 84
- wavelengths of electrons 82
RHO 10, 158, 159
- conformation of 158
- conformational changes in 158
- framework of 10
- prediction of 159
- topology of 10
Rietveld method 36, 37
Rietveld refinement 66, 128, 130
Rietveld technique 6
rigid unit modes (RUMs) 135
RUB-10 24
RUB-13 24
RUB-17 (RNS) 24
- end members of 24

RUB-n 23, 24
- end members of 24
- faulting in 23
RUM 135

S
SAPO-31 (ATO) 25
- isomorphous substitution in 25
- X-ray diffraction data of 25
SAPO-40 (AFR) 6, 18
- template ordering in 18
satellite reflections 129, 134
SBUs 151, 152
scanning electron microscope (SEM) 75
- high spatial resolution 75
- schematic diagram 75
- surface topology 75
scanning electron microscopy (SEM) 72, 79, 186
- of natural zeolites 79
scattering amplitudes 74
scattering cross section 33, 34
- coherent 34
- incoherent 34
scattering factors 34
- neutron 34
- X-ray 34
scattering length 34
- depending on the nuclear spin state 34
scattering vector 73
Scherzer focus 83
- CTF 83
scoletite 59
- hydrogen bond scheme 59
scoletite, $CaAl_2Si_3O_{10} \cdot 3\,H_2O$ 58
- Al/Si ordering in 58
- neutron diffraction by 58
- water molecules in 58
sealed glazing units 193
second coordination networks 156
secondary building units (SBUs) 150, 153
seeding techniques 173
seeds 172
- influence of 172
SEM 76, 78, 83, 84
- crystallinity 78
- field emission guns (FEG) 83
- morphology 78
- resolution 83
- sample preparation for 84
- schematic diagram 76
- stability of crystals under an electron beam 78
- symmetry 78
SEM (HRSEM) 85
- charging problems 85
- high resolution 85
- low energy electrons 85
sepiolite 181, 183
shear strain 120
shearing 124, 131
- in sodalite frameworks 124
Si–O bond length 118
Si–O–Si angle 118
Si/Al ratio 97
- by HRTEM images 97
- modulation of 97
- in MOR 97
silica sources 169, 178
silicalite-1 180
silicoaluminophosphate 64
- protons in 64
silicone foam 180
- containing zeolite A powder 180
silicone rubber 195
- loaded with molecular sieve powder 195
simulated ED patterns of MEL 91
single-crystal data 2, 5, 8
- accuracy of 2
- determination of the unit cell 5
- of gismondine 8
- reflections 5
- symmetry 5
single-crystal neutron diffraction 32
- absence of peak overlap 32
single-crystal studies 66
sodalite (SOD) 10, 144
- danalite 144
- frameworks collapse of 10
- hauyn 144
- nosean 144
- powder diffraction of 10
- space group symmetry of 10
- topology of 10
- tugtupite 144
sodalites 127, 128, 130
- modulations in 128, 130
- thermal expansion of 127
sodium aluminate 170, 175
sodium disilicate 170
sodium water glass 170, 175
sodium zinc arsenate, $Na_6(ZnAsO_4)_6$ 46
- neutron diffraction of 46
- reversible transformation on rehydration of 46
- structure of 46
- X-ray powder diffraction by 46
soil conditioning 166
sorbate molecules 22

- by single-crystal X-ray diffraction 22
- disorder of 22
- symmetry of 22
space group 5, 6
- determination of 5
- maximum symmetry of the topology 5
- by powder data 5
- starting topology for 6
space group types 118
- three-dimensional 118
spherical aberration 81
SSZ/CIT series (CON) 24
- end members of 24
SSZ-26 100
- ED patterns of 100
- framework structures of 100
- HRTEM images of 100
- X-ray powder diffraction of 100
SSZ-33 100
- ED patterns of 100
- framework structures of 100
- HRTEM images of 100
SSZ-n/CIT-n series 23
- faulting in 23
stacking faults 155
starting models 26
- for molecular modeling techniques 26
- space group for 26
- unit cell dimensions for 26
starting topology 6
static RUMs (rigid unit modes) 135
- linear combinations of 135
stellerite $Ca_{7.7}Na_{0.3}(Al, Si)_{72}O_{144} \cdot 50\ H_2O$ 60
- structure of 60
stilbite 60, 144
- barrerite 144
- framework structure of 60
- neutron diffraction by 60
- stellerite 144
stirrer speed 177
- during synthesis 177
stirring speed 172
string constants 131
structural distortions 113
- in microporous materials 113
structural modulations 128
structure 6
- space groups 6
structure determination 99, 107
- by electron dynamic scattering 107
structure determinations 25
- of unknown materials 25
- starting models for 25
structure directing agent 17
- triple helix 17
structure factor 36
structure refinement 37
structure solution 5
- space groups 5
structure type codes 145, 146, 151, 152
- isotypes 145, 146
- table 145, 146
- type material 145, 146
structure types (ST) 142-144, 147, 150, 154
- associated data 144
- classification by Breck 154
- classification by Gottardi and Galli 154
- classification system 154
- codes 144, 151, 152
- isotypes 144
- largest aperture 147
- number of 143
- smallest loops 150
- table of 144
structures 128
- modulated 128
superspace 129
superspace groups 118, 129
superstructures 136
- commensurate 136
surface structure 92
- chemical reaction 92
- crystal growth mechanism 92
sweetening 195
symmetry 8, 118, 119
- changes upon dehydration 8
- of a zeolite framework 118
- real 119
- topological 119
symmetry changes 11
- number of reflections 11
synchrotron X-ray data 3
synchrotron X-ray sources 25
- single-crystal data 25

T
T-O bond lengths 117
T-atoms 141, 144
- Si, Al, P 141
- Si, Al, Ga, Ge, B, Be, Zn, Co etc. 144
TEM 84
- sample preparation for 84
TEM images 76
- amplitude contrast (diffraction) 76
- intensity contrast (absorption) 76
- phase contrast (out-of-focus) 76
template 180
- cost of 180
template ordering 18

template
- cobalticinium template cation 18
- morpholinium template 18
- powder data of 18
- single crystal data of 18
tetraalkoxysilanes 169
tetragonal tetrahedron distortion 119
tetrahedra 126
- isolated 126
tetrahedron distortion 114, 119, 121, 126
- cation contribution to 121
- framework contribution to 121
- systematic 119
tetramethylammonium sodalite 126
theoretical networks 160
- graph theory 160
thermal expansion 126, 135
- negative 135
- of tetrahedral framework structures 126
thermal swing adsorption (TSA) 165
thomsonite $NaCa_2Al_5Si_5O_{20} \cdot 6\,H_2O$ 60
- neutron diffraction of 60
- neutron refinement for 60
- Si/Al ordering in 60
- water molecules in 60
- X-ray diffraction by 60
three-ring opening 6
tilt angle 122, 123, 131
tilt systems 123, 136
time-of-flight method (TOF) 35
titanosilicate 100
toluene 53
- adsorption sites by molecular mechanics calculations 53
topological density (TD) 157
topological search routine 5
topological symmetry 5, 118
topology 142
topology-induced stress 119
TOT angles 117
TOT bridges 118
- straight 118
transmission electron microscope (TEM) 75
- back-focal plane 75
- image plane 75
transmission EM (TEM) 72
transmission function $\phi(x, y)$ 79, 81
triple helix 7, 43
- occluded water 7
- water molecules in 43
triple water helix 17
- in VPI-5 17
Tschernichite 23
TSF process 197
twisting 126
- local 126

U
unfolding 127
- of the framework 127
unit cell 3, 9
- determination of 3
- maximum volume of 9

V
vacuum pressure swing adsorption (VPSA) 197
vitrification 104, 105
- radiolytic process 105
VPI-5 (VFI) 7, 8, 17, 18, 43
- di-n-propylamine for synthesis of 18
- octahedral aluminum in 7
- reconstructive phase transition of 7
- refinement of the structure 7
- structure refined from neutron powder data 43
- symmetry of 7
- synthesis of 18
- transition of 8
- triple helix of water in 17
- water molecules in 43
VPI-7 (VSV) 24
- end members of 24
VPI-9 4, 5
- complex framework of 5
- structure determination of 4

W
water glass 169
water molecules 63
- coordinated to cations 63
- H–O–H angle 63
- hydrogen bonded to framework 63
- O–H distances 63
wave field, $\Phi(x, y)$ 81

X
X (FAU) 18
- organic molecules adsorbed into 18
X-ray diffraction 1, 66, 142
- microcrystals 142
- refinements 66
- structure determination 1
- synchrotron radiation for 142
X-ray fluorescence spectroscopy (XRF) 186
X-ray powder diffraction 184
X-rays 74
- diffraction behavior 74
XRD powder data 143

- number of structure types 143
xylene 50, 52
- in Na-X 52
- in Yb, Na-Y 50
- modification of the unit-cell parameter by 50
xylene isomers 50
- in Yb, Na-Y 50
xylenes 50, 53
- adsorption sites by molecular mechanics calculations 53
- in FAU structures 50
- neutron diffraction of 50

Y
Y (FAU) 18
- aniline in 18
- *m*-dinitrobenzene in 18
- mesitylene in 18
- organic molecules adsorbed into 18
Y-zeolites 64
- hydrogen positions in 64
yugawaralite, $CaAl_2Si_6O_{16} \cdot 4\,H_2O$ 61
- neutron diffraction by 61
- ordered Si/Al in 61
- water molecules in 61

Z
Z-contrast method 94
- localization of particles by 94
ZAPO-M1 (ZON) 25
- structure of 25
- X-ray diffraction data of 25
zeolite A, 3A, 4A, 5A (LTA) 15, 16, 22, 121, 125, 127, 165, 174–176, 178, 180, 184, 188, 192
- bulk density of 188
- calcium-exchanged 121
- crystallization field for 178
- demand for 165
- effective size of the pore opening 16
- exchanged with Pb^{2+} 22
- framework flexibility of 15
- hydrothermal stability of 192
- in detergents 165
- industrial synthesis of 174
- interatomic distances in 16
- non-collapsible framework of 15
- powder structural data of 15
- single-crystal structural data of 15
- single-crystal structure refinements 16
- SEM photograph of 187
- symmetry changes of 16
- synthesis of 176
- thermal expansion of 127
- X-ray diffraction pattern of 184
zeolite A powders 180
- in the polyurethane industry 180
zeolite beta (BEA) 99
- ED pattern of 99
- electron microscopy of 99
- HRTEM images of 99
- structural faulting of 99
zeolite/cation systems 22
zeolite crystals 155
- structural faults of 155
zeolite D-Rho 41
- dealumination of 41
- flexibility of the framework of 41
- non-framework aluminum in 41
zeolite D-ZK-5 41
- dealumination of 41
- deep-bed calcined 41
- IR of 41
zeolite frameworks 115
- neutron diffraction by 41
- NMR of 41
zeolite L 54
- benzene in 54
zeolite materials 104
- electron beam sensitivity of 104
zeolite Rho 41, 42, 127, 128
- Cs-exchanged 127
- flexibility of 42
- flexibility of the framework of 41
- framework of 42
- multiphase Rietveld refinement of 41
- neutron diffraction by 41
- Sr- and NH_4-exchanged 128
- two coexistent phases of 41
zeolite stability 191
zeolite structure types 148, 153
- classification of 153
- phosphates 148
- silicates 148
zeolite uses 164
zeolite X 177–179, 195
- crystallization field for 179
- in the purification of air 195
- industrial synthesis of 177
- low SiO_2/Al_2O_3 of 178
zeolite Y 40
- adsorbed benzene molecules in 40
- of a highly crystalline "zero defect" nature 40
zeolite ZK-4 40
- framework ordering in 40
- structure of 40
zeolite ZSM-5 179
- industrial synthesis of 179

zeolite ZSM-5
- SEM photograph of 187
zeolite-like materials 148
- framework density (FD) of 148
zeolites 78, 148, 166
- criterion 148
- electron beam sensitivity of 78
- framework density (FD) of 148
- industrial synthesis of 166
zeolites for adsorption processes 183
- characterization of 183
zeolite-type framework structures 141, 144, 147
- apertures of 147
- channel systems of 147
- connectivity 144
- ring openings of 147
- topology 144
zincophosphate (CZP) 17
- chiral 17
zincosilicates 150
- 3-rings in 150
ZSM-3 155
ZSM-5 (MFI) 11, 109
- containing the tetrapropylammonium (TPA) ion 11
- ED pattern of 109
- HRTEM image of 109
- presence of TPA in 109
- structure of 11
ZSM-18 (MEI) 6
ZSM-20 155

Author Index Volumes 1 and 2

Baerlocher C see Meier WM (1999) *2*:141 - 161
Beck JS see Vartuli JC (1998) *1*:97 - 119
Bellussi G see Perego G (1998) *1*:187 - 228
Bennett JM see Koningsveld van W (1999) *2*:1 - 29

Coker EN, Jansen JC (1998) Approaches for the Synthesis of Ultra-Large and Ultra-Small Zeolite Crystals. *1*:121 - 155
Cool P, Vansant EF (1998) Pillared Clays: Preparation, Characterization and Applications. *1*:265 - 288

Depmeier W (1999) Structural Distortions and Modulations in Microporous Materials. *2*:113 - 140

Ernst S (1998) Synthesis of More Recent Aluminosilicates with a Potential in Catalysis and Adsorption. *1*:65 - 96

Fitch AN, Jobic H (1999) Structural Information from Neutron Diffraction. *2*:31 - 70

Gies H, Marler B, Werthman U (1998) Synthesis of Porosils: Crystalline Nanoporous Silicas with Cage- and Channel-Like Void Structures. *1*:35 - 64

Jansen JC see Coker EN (1998) *1*:121 - 155
Jobic H see Fitch AN (1999) *2*:31 - 70

Kresge CT see Vartuli JC (1998) *1*:97 - 119
Koningsveld van W, Bennett JM (1999) Zeolite Structure Determination from X-Ray Diffraction. *2*:1 - 29

Marler B see Gies H (1998) *1*:35 - 64
McCullen SB see Vartuli JC (1998) *1*:97 - 119
Meier WM, Baerlocher C (1999) Zeolite Type Frameworks: Connectivities, Configurations and Conformations. *2*:141 - 161
Millini R see Perego G (1998) *1*:187 - 228

Perego G, Millini R, Bellussi G (1998) Synthesis and Characterization of Molecular Sieves Containing Transition Metals in the Framework. *1*:187 - 228
Pfenninger A (1999) Manufacture and Use of Zeolites for Adsorptions Processes. *2*:163–198

Roth WJ see Vartuli JC (1998) *1*:97 - 119

Schüth F see Schunk SA (1998) *1*:229 - 263
Schunk SA, Schüth F (1998) Synthesis of Zeolite-Like Inorganic Compounds. *1*:229 - 263
Szostak R (1998) Synthesis of Molecular Sieve Phosphates. *1*:157 - 185

Terasaki O (1999) Electron Microscopy Studies in Molecular Sieve Science. *2*:71–112
Thompsen RW (1998) Recent Advances in the Understanding of Zeolite Synthesis. *1*:1–33

Vansant EF see Cool P (1998) *1*:265–288
Vartuli JC, Roth WJ, Beck JS, McCullen SB, Kresge CT (1998) The Synthesis and Properties of M41S and Related Mesoporous Materials. *1*:97–119

Werthmann U see Gies H (1998) *1*:35–64

Springer and the environment

At Springer we firmly believe that an international science publisher has a special obligation to the environment, and our corporate policies consistently reflect this conviction.
We also expect our business partners – paper mills, printers, packaging manufacturers, etc. – to commit themselves to using materials and production processes that do not harm the environment. The paper in this book is made from low- or no-chlorine pulp and is acid free, in conformance with international standards for paper permanency.

MIX
Papier aus verantwortungsvollen Quellen
Paper from responsible sources
FSC® C105338

If you have any concerns about our products,
you can contact us on
ProductSafety@springernature.com

In case Publisher is established outside the EU,
the EU authorized representative is:
Springer Nature Customer Service Center GmbH
Europaplatz 3, 69115 Heidelberg, Germany

Printed by Libri Plureos GmbH
in Hamburg, Germany